Bin Tian

Wireless Communications

Information and Computer Engineering

Volume 9

Bin Tian

Wireless Communications

—

DE GRUYTER SCIENCE PRESS

Author
Bin Tian
School of Telecommunications Engineering
Xidian University
btian@xidian.edu.cn

ISBN 978-3-11-075135-2
e-ISBN (PDF) 978-3-11-075143-7
e-ISBN (EPUB) 978-3-11-075152-9
ISSN 2570-1614

Library of Congress Control Number: 2023949971

Bibliographic information published by the Deutsche Nationalbibliothek
The Deutsche Nationalbibliothek lists this publication in the Deutsche Nationalbibliografie; detailed bibliographic data are available on the Internet at http://dnb.dnb.de.

Cover image: Prill/iStock/Getty Images Plus
Typesetting: Integra Software Services Pvt. Ltd.
Printing and binding: CPI books GmbH, Leck

www.degruyter.com

Author biography

Bin Tian received his Bachelor degree, Master degree and Ph.D. degree from Xidian University, Xi'an, China, in 1992, 1995 and 2000, respectively. He is a professor at the School of Telecommunications Engineering, Xidian University.

https://doi.org/10.1515/9783110751437-202

Preface

Wireless communication is a dynamic and vibrant area in the communication industry in the past decades due to several factors. Firstly, there has been an explosive increase in demand for wireless access, driven by cellular system and wireless data applications. Secondly, the dramatic progress in signal processing algorithms and VLSI technology has enabled small-area and low-power implementation of sophisticated processing and coding techniques. Thirdly, the successful standardization of wireless systems, such as the second-generation (2G) standards (GSM and IS-95), the third-generation (3G) standards (WCDMA, CDMA2000, TD-SCDMA and WiMAX), the fourth-generation (4G) standards (TD-LTE), and fifth generation (5G) under development, provides a concrete demonstration that good idea from wireless communications can have a significant impact on life. People have devoted great enthusiasm to wireless communication technology. Wireless communication has great economic benefits. It will produce very broad application prospects and will greatly change the way of people's learning, working and living. Wireless communication systems and networks are being installed everywhere throughout the world. People at any location use handheld wireless devices and laptops to remain connected to other people and information source.

Thus, *wireless communications* become an important course for students and engineers specialized in telecommunication engineering. My academic profession is in wireless communications, and I have been teaching the course *wireless communications* in English since 2004. This course was rated as national bilingual teaching demonstration course in 2010. During the course of teaching, I strongly felt that there was a need of a comprehensive and brief textbook which deals with all aspects of wireless communications and which can lay a solid foundation for further academic research and engineering career. Based on my experience in research and teaching, I was motivated to share the academic and industries' needs through a comprehensive but concise enough and easy-to-understand textbook. We assume the reader has basic knowledge of the course of *principle of communications*; the fundamental techniques of communications such as modulation, channel coding and synchronization have not been discussed in detail and only used for the discussion for the performance of communications systems over wireless channels. We focus on the wireless channel, the problems it might bring and techniques to overcome them. We also give an example to show how a simple cellular communication system works. This book is primarily written as a textbook for postgraduate students, and it is also used as a book for engineers major in Electronics and Telecommunications Engineering.

The chapter and material sequence in this textbook have been designed mainly for postgraduate-level course. Chapter 1 gives an overview of the wireless communications and includes the history, the technical challenges, wireless systems and standards. Chapters 2 and 3 discuss the wireless channels. Chapter 2 deals with the large-scale path loss and shadowing, while Chapter 3 presents the small-scale fading and

https://doi.org/10.1515/9783110751437-203

statistical multipath channel. In Chapter 4, we discuss the digital modulations and their performance over wireless channels. Chapters 5 to 7 discuss the techniques used in wireless systems to overcome the problems caused by fading. Chapter 5 discusses diversity, channel coding and equalization. In Chapter 6, we discuss spread spectrum systems and spread spectrum with RAKE receiver. Chapter 7 deals with multicarrier modulation and multi-antenna systems. In Chapter 8, we discuss techniques necessary to build a cellular system, such as duplexing, multiple access and cellular design fundamentals. We also include a simplified cellular system to ensure the reader understand how a cellular system works. We believe that the way of writing this book is helpful for readers to understand wireless communication more quickly and simply. We have not given the standard wireless systems, which involve more concepts and more factors resulting from the balance of interests in the formation of standards. However, we still recommend that readers pay attention to the latest information in the wireless communication industry and academia, the specific standards of wireless systems, and even the progress of new standards.

Bin Tian
Xi'an, China

Acknowledgments

I wish to express my sincere thanks to the reviewer of this book. The manuscript was reviewed by Prof. Huaxi Gu, Prof. Guangliang Ren and Dr. Jie Guo at the School of Telecommunications Engineering. Constructive suggestions and corrections were given. The manuscript was also reviewed by Prof. Yang Yao at the School of Foreign Languages for language problems. I would like to thank Prof. Xu Wang at Heriot-Watt University for his valuable suggestions.

I also thank my students Sen Zhang, Senhao Zhang, Xiaojuan Du, Jingwu Lyu, Hong Zhao and Yifan Dai for their assistance in drawing the figures, make the tables and checking the format of the full text.

In writing this textbook, materials from many excellent published books, as well as papers from journals, magazines and conferences were used. I would like to express my deep gratitude to the authors, and these publications will be included in the reference.

During the writing the book, my family have provided infinite support and love which is the key factor to finish the book. My wife has been always supporting me for everything in my career life including the writing of the book. She has made many sacrifices and allowed me to spend more time on my work and this facilitated the smooth completion of this book. This book is dedicated to her.

https://doi.org/10.1515/9783110751437-204

Contents

Chapter 1
Wireless communication overview

History has proven that wireless communication has already changed the way people communicate with each other. With the progress of techniques related wireless communications, many innovative ways of communication have become a reality, thus making it possible for people to communicate with anyone or any device at any time or any place. Nowadays, wireless communication has captured the attention and the imagination of the media as well as the public. Especially the 5G mobile communication system under development and the international trade disputes related to 5G systems bring this field of professional technology to the ordinary people, which arouse people's great enthusiasm and infinite yearning for high-performance wireless communications. The most typical examples are the exponentially growing cellular phones and the ubiquitous WiFi coverage. However, what is more exciting is that we continue to see new systems and applications, such as wireless sensor networks, automated highways and factories, driverless cars and unmanned stores, smart homes and appliances, and remote telemedicine, are emerging from research ideas or laboratories to concrete systems.

The explosive growth of wireless systems coupled with the proliferation of laptop and other smart terminals indicate a bright future for wireless networks. We will first review the history of wireless networks in this introductory chapter and then discuss the technical challenges. We also describe the current systems and standards.

1.1 History of wireless communications

1.1.1 Wireless communication in preindustrial period

Nobody in this world knows exactly when the history of wireless communications started. The first wireless communication networks started in the Pre-industrial age. The famous Chinese story that the last ruler of the Western Zhou Dynasty, King You, made fun of the feudal lords by using beacon-fire took place at 779 B.C. is a kind of typical wireless communications by using fire and smoke signals. These wireless systems transmitted information over line-of-sight (LOS) distances by using some simple signals, such as smoke signals (in daytime), fire signal (in the night), signal flares or beating drums. These signals can also be relayed by observation stations such as beacon tower to large distances. When telegraph and telephone were invented, these modes of communications have become the thing of the past.

https://doi.org/10.1515/9783110751437-001

1.1.2 First radio transmission demonstration

The first radio transmission was demonstrated by Tesla in 1894, not long after Maxwell and Hertz laid the foundation for our understanding of the transmission of electromagnetic waves. However, people generally recognize the transmission by Marconi as the first radio transmission. The famous radio transmission from the Isle of Wight to a tugboat 18 miles away was demonstrated by Marconi in 1898. Marconi won the Nobel Prize in 1909 as the inventor of wireless communications. However, there has been a lot of controversy about who invented radio transmissions. A couple of people that were famous, including Edison and Karl Ferdinand Braun, claimed to have invented radio first. On June 21, 1943, the Supreme Court of the United States ruled that the inventor of the radio patent was Nicolas Tesla. It should be mentioned that wireless radio transmission system started quite early in China. In 1899, wireless radio station was set up in Guangzhou viceroy office, fortresses such as Makou and Qianshan, and larger ships by the governor-general of Guangdong and Guangxi provinces.

1.1.3 Wireless voice transmission and the cellular systems

The first wireless voice transmission began in 1915, in which the transmission between New York and Francisco was established. The first bi-directional wireless systems emerged in 1930s. These systems were used in police departments and the military within closed user group. During and after the Second World War, many researches were driven by military applications. During 1940s and 1950s, there are several important developments. In 1946, the first wireless mobile system was installed in USA. However, the interface between this system and the PSTN (Public Switched Telephone Network) was different from the interface nowadays. It was not automated at that time, and it consisted of human telephone operators. After that, German (in 1950), French (in 1956) and the UK (in 1959) established their public mobile telephone systems. These systems used one very tall transmitter to cover an entire big city. The use of radio spectrum is very inefficient at that time, and even in 1970s, the mobile telephone network in New York could only support 543 subscribers.

To solve this capacity problem emerged in the 1950s and 1960s, the cellular concept was developed by the researchers at AT&T Bell Laboratories (MacDonald 1979). Today, the cellular telephone system has been the most successful application of wireless networking. The cellular concept utilizes the path loss characteristics that the power falls off quickly with distance. In this way, two subscribes can operate on the same frequency band with enough distance so that the signal-to-interference power ratios (SIR) is good enough. In this way, more subscribers can be supported with the same frequency band. The first cellular system is analog and was developed in Chicago in 1983. Only 1 year later, FCC had to authorize 10 MHz more bands to support more subscribers. The analog system grew rapidly in many countries. For example, during 1980s, it reached market penetration of up to 10% in Europe. In China, the first analog cellular mobile phone system was established in 1987 with 700 users.

Only the first generation of cellular system is the analogy system. The second generation of cellular systems was all based on digital communications, which was developed in early 1990s. There were mainly two standards in the second-generation cellular systems, GSM and IS-95, which was developed in Europe and USA separately. There is only one standard used in Japan. It first supports mainly voice and then gradually supports data services such as short messaging, e-mail and Internet access.

The third-generation cellular communication systems were deployed around 2000. They provide higher data rates from 144 kbps to more than 2 Mbps at speed up to 500 km/h. There are three main standards: WCDMA, CDMA2000 and TD-SCDMA, proposed separately from Europe, USA and China, it should be noticed that they are not compatible.

The fourth-generation cellular systems have been deployed since 2012. There are two standards: TD-LTE and FDD-LTE. Of course, strictly speaking, only the upgraded LTE-Advanced can meet the requirements of the International Telecommunication Union for 4G. The transmission rate of the fourth-generation cellular system can reach 20 Mbps or even up to 100 Mbps.

The fifth-generation wireless system is currently the latest generation of cellular mobile communication system. It was originally planned to be deployed in 2020, but in fact in 2019, it has begun to be deployed. The performance goal of 5G is to increase data rate, reduce delay, save energy, reduce cost, increase system capacity and connect large-scale equipment. The International Telecommunications Union (ITU) IMT-2020 specification requires a speed of up to 20 Gbps, which requires wide channel bandwidth and massive MIMO.

1.1.4 Packet radio

In the history of wireless communications, packet radio played a very important role. It belongs to digital radio communications, and the bits are grouped into packets to be sent in burst. The first packet radio network is ALOHANET which is developed at the University of Hawaii in 1971. This network enabled computers at seven campuses spreading out over four islands to communicate through the central site computer on Oahu, which is the hub. It should be mentioned that the protocols used in ALOHANET laid the foundation for packet radio protocols. The fundamental of combined packet data in broadcast radio attracted the US military. In the following two decades, the US army made great efforts in developing wireless packet networks for the battlefield. The network is called ad hoc network because the nodes can be reconfigured without pre-established infrastructure. However, the networks at that time were poor in performance. Besides in military scenario, it does not succeed in commercial uses. One reason is probably that the wired Ethernet is so good that it goes far beyond wireless network.

There was a turning point in 1985. The Industrial, Scientific and Medical (ISM) frequency bands were allowed in commercial uses for the development of wireless LANs by the Federal Communications Commission (FCC) at that year. It was not necessary for the wireless LAN vendors to obtain an FCC license to develop WLANs in this band. However, there was a drawback. In order not to interfere with the primary ISM band devices, the WLAN devices had to use low-power density schemes. It also needs techniques to overcome the interference from primary subscribers. As a result, the first-generation WLAN systems have low data rates and short coverage. Furthermore, there is not a standard for the initial WLANs so that they did not attract much attention.

The IEEE standard 802.11b adopted in 1999 became the standard for the second-generation WLAN, and second-generation WLANs achieve a data rate of 11 Mbps and coverage size is about 100 m. The IEEE 802.11a wireless LAN standard is another WLAN standard, and it is based on multicarrier modulation. It is also adopted in 1999 and has the data rates of 20–70 Mbps. It has a frequency band of 300 MHz in the 5 GHz U-NII band. IEEE 802.11g that supports WiFi services is adopted in 2002. Its frequency band can be either the 2.4 GHz or the 5 GHz, and the data rate is up to 54 Mbps. Similarly, multicarrier modulation is adopted. The 802.11n, adopted in 2009, provides 300–600 Mbps. It adopts MIMO-OFDM technology, improves the quality of the wireless transmission and makes the transmission rate a great ascension. In coverage, the 802.11n adopts the smart antenna technology. And it can dynamically adjust the beam forming to ensure the WLAN users receive the stable signal. Therefore, its coverage can expand into several square kilometers, and this greatly improves the mobility of the WLAN. In the compatibility, 802.11n adopted a software defined radio technology, it is a fully programmable hardware platform and makes different system stations and terminals can all work through this platform software, which makes the interchange and compatibility of WLAN a great improvement.

1.1.5 Radio paging

Radio paging systems are another example of wireless data network, used to be very successful before the development of the second-generation cellular system that support short messaging services (SMS). These systems can be regarded as one-way mobile communication. The pager is actually a receiver and it is used to receive a paging message from the paging center. In 1948, Bell Labs developed the world's first pager, named Bell Boy. After entering the 1970s, by using new technologies, it gained great success, with 50 million subscribers in the US alone. Pagers and wireless paging networks were introduced to China in 1980s. After 10 years of development, the paging system reached its peak, with 65 million subscribers in 1998, in China. However, these systems were starting to wane with the popularity of cellular telephone systems. Paging systems allow coverage over very wide areas by simultaneously broadcasting the pager message at high power from multiple base stations (BSs) or satellites. These systems have been around for many years. Early radio paging systems were analog 1 bit messages to inform a user that someone was trying to reach him or her. The caller needs callback by using the regular telephone system to obtain the phone number of the paging party. Lately, the paging systems allowed a short digital message, which includes a phone number and brief text, to be sent to the pager receiver. In paging systems most of the complexity is built into the transmitters so that pager receivers are small and lightweight and have a long battery life. The network protocols are also very simple since broadcasting a message over all BSs requires no routing or handoff.

1.1.6 Satellite communications

Commercial satellite communication systems are now emerging as another major wireless communications. Satellite systems can provide broadcast services over very wide areas and are also necessary to the coverage gap between high-density user locations. Satellite mobile communication systems follow the same basic principle as cellular systems, except that the cell BSs are now satellites orbiting the earth. Satellite systems can be classified three categories by the height of the satellite orbit: low-earth orbit (LEO) where the altitude is around roughly 2,000 km, medium-earth orbit (MEO) where the altitude is around roughly 9,000 km, and geostationary Earth orbit (GEO) where the altitude is around roughly 36,000 km. The GEO satellite's operating period is equal to the earth's rotation period (23 h 56 min and 4 s) and runs in the same direction as the Earth's rotation so that they are seen as stationary from the earth. The disadvantage of high altitude orbits is that the signal path-loss is high, and the propagation delay is too big for the real-time application such as voice. The advantage is that the satellites can cover a wide range. The idea of covering the world with three GEO satellites was first proposed by the science fiction writer Arthur C. Clarke in 1945. However, the first two satellite systems in deployed in the cold war period, Sputnik by the Soviet Union's in 1957 and the Echo-1 deployed by the US NASA/Bell Laboratories in 1960 were not GEO satellites. It was a very difficult task to lift such a huge object to 36,000 km height. Until 1963, the Hughes and NASA of USA launched a GEO satellite, pulling back a round in the cold war. In 1964, the International Telecommunications Satellite Organization (INTELSAT) was established. The first commercial synchronous satellite, *Early Bird* INTELSAT-1, was successfully launched in 1965. In 1970, China launched its first satellite, DFH-1. Early satellite systems were mostly based on GEO satellites. For example, INMARSAT is the most well-known GEO satellite system still in operation today. Later, LEO satellites were focused on so that they could support higher rate transmission with a lower delay. Despite some important failures, such as the failure of the Iridium satellite system, there is no way to stop people from launching satellites. Table 1.1 shows the number of satellites in orbit in different countries by the end of 2018.

Tab. 1.1: Number of satellites in orbit by the end of 2018.

USA	865
China	289
Russia	150
Japan	73
India	56
UK	54
Other countries and organizations	531

1.2 Technical challenges of wireless communications

In designing a wireless communications system, there are many technical problems that must be solved. At the hardware level the terminal must have multiple modes of operation to support the different applications and media. Whether it's a desktop computer, a handheld computer, a mobile phone, a watch or other wearable device, it is hoped that it can support a variety of service such as voice, image, text and video data. Most people don't want to carry around 1 kg battery, the signal processing and communications hardware of the portable terminal must consume very little power, which will impact higher levels of the system design (Boccuzzi 2008). The main challenges to wireless communications system in high level can be summarized as the signal propagation characteristics and fading effects, the spectrum limitations, the limitations in energy constrains, the user mobility and the cross-layer design issues (Molisch 2011, Goldsmith 2005).

1.2.1 Signal propagation characteristics and fading effects

For wireless communications, the transmission link between the transmitter and receiver is the wireless channel. In free space, the received power falls off with the square of the distance, d^{-2}, where d is the transmitter and receiver separation distance. In Chapter 2, we will see that the power generally decrease with distance much faster than d^{-2} because of the presence of other disturbances in the environment. In the actual environment, there are usually several obstacles between the transmitter and the receiver, and further, the obstacles always absorb some power while scattering the rest. Experimental field studies show that power decay is like d^{-2} when the receiver is near the transmitter, and it decays exponentially with distance $d^{-\gamma}$, where γ range from 3 to 8 at large distances.

When there are obstacles between the transmitter and receiver, the wave (line of sight (LOS) or other multipath) go through or around the obstacles is greatly attenuated. This effect is called *shadowing*, which is shown in Fig. 1.1. Because of diffraction effect and spatially extended secondary radiation sources, the transition from "light" (LOS) zone to the "dark" (shadowed) zone is gradual. The terminal needs to take a large movement (from a few meters to several hundred meters) to move from large to dark zone. For this reason shadow is *large-scale fading*. Since variations due to path loss and shadowing occur over relatively large distances, this variation is sometimes referred to as *large-scale propagation effects.*

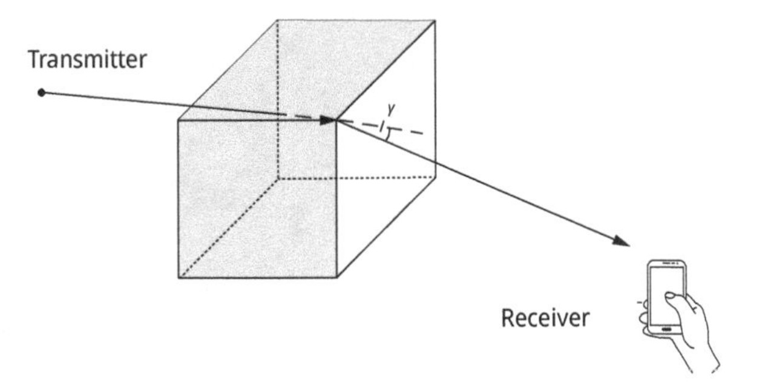

Fig. 1.1: Shadowing by a building.

The signal is transmitted over different propagation paths, as shown in Fig. 1.2. In some cases, a LOS connection might exist between the transmitter and receiver. Moreover, the signal can get to the receiver by being reflected, diffracted or scattered by the environment: houses, mountains, windows, or walls. The number of these possible paths generally is very large, random and varying with time. Each of the paths has a distinct amplitude, delay (running time of the signal), direction of departure from the transmitter, direction of arrival and phase shift. Most importantly, the number of the paths and the amplitude, delay, phase and the Doppler shift of each path may be changing all the time due to movement of transmitter, receiver and/or the surrounding objects. Because of the changing reflection, diffraction and scattering, the signal is actually a random process. These effects, namely, the changing of the total signal amplitude due to interference of the different multipath components is called small-scale-fading or small-scale propagation effects. A small movement such as 10 cm at the order of wavelength can affect a change between destructive and constructive interference, which result in large change in signal amplitude. The path loss property and fading make the received power of the signal unpredictable, which makes the development and deployment of the wireless system difficult.

The difference in path distance leads to the different runtimes for different multipaths. This leads to different phases of multipaths and will result in interference in narrowband systems. However, when the signal does not satisfy the narrow band criterion, the multipath delay spread will cause another form of distortion. In this scenario, we assume the multipath delay spread is T_m, a short-transmitted pulse with a duration of T will result in a received signal with a duration of $T + T_m$. In this way, the duration of the received signal might be increased significantly. This situation is illustrated in Fig. 1.3. This figure shows a pulse of width T that is transmitted over a multipath channel. We will discuss in Chapter 4 that linear modulation consists of a train of pulses, and each pulse carries information in its amplitude or phase which corre-

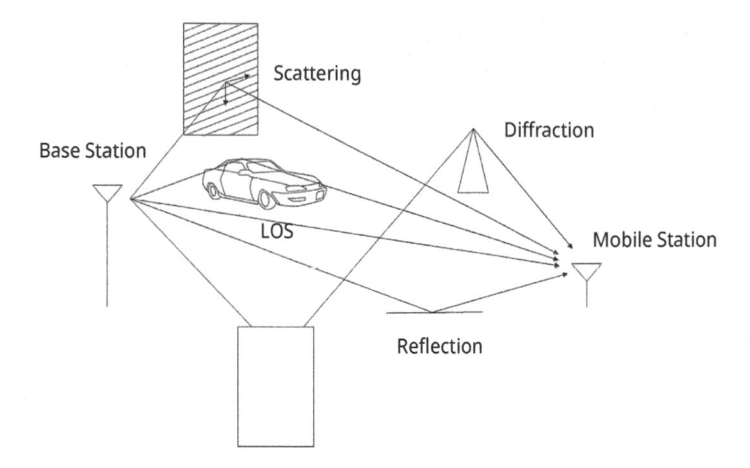

Fig. 1.2: Multipath effects caused by various obstacles.

sponds to a data bit or symbol. If the multipath delay spread is much great than the symbol time, the subsequently transmitted pulses will be interfered with by these multipath components. This effect is called intersymbol interference (ISI).

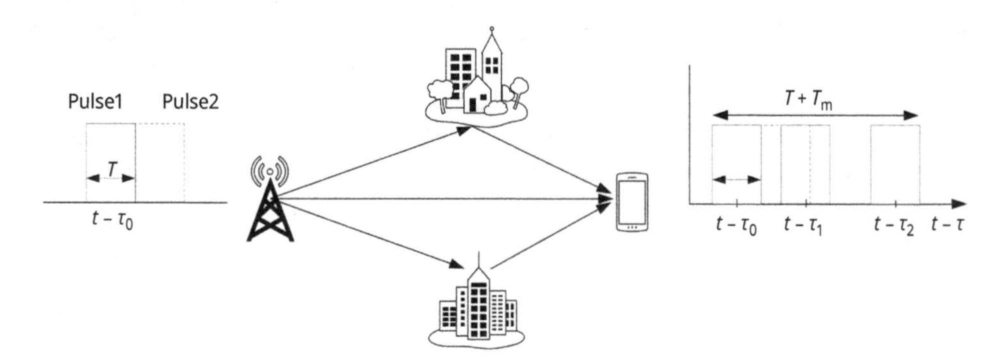

Fig. 1.3: ISI effects.

1.2.2 Spectrum regulations and limitations

Spectrum is a public resource and a limited resource. Its use is constrained and regulated by international agreements, allocated and controlled by government agencies in many countries. So it must be used in a highly efficient way. There are two main kinds of spectrums: regulated spectrum usage, where the spectrum is authorized by a certain agency to and controlled by a single network operator, and unregulated spectrum usage, where it is unnecessary to get a license, but the use of it needs to satisfy the restriction on transmit power and bandwidth, and specified services.

1.2.2.1 Regulated spectrum

The Federal Communications Commission (FCC), the European Telecommunications Standards Institute (ETSI), the China National Radio Administration (CNRA) and the Association of Radio Industries and Business (ARIB) are responsible for the commercial spectral allocation in the United States, Europe, China and Japan, respectively. The Office of Spectral Management is responsible for the spectral allocation of military use in USA. Globally, the commercial spectral allocation is regulated by ITU. The exact frequency assignments and assignment policy are different in different countries and regions. For example, in USA, FCC generally allocate spectral blocks for specific purposes based on a number of criteria historically, but now it is common to assign the block trough spectral auction to the highest bidder. Some countries use marked-based policy, some use unified planning approach. Table 1.2 shows the major licensed bands for broadcasting and communications. This is quite similar all over the world. We give the 5G spectrum allocation for several countries to show the difference, you may refer to the corresponding country's radio frequency (RF) allocation regulations if you are interested in the frequency allocation of a specific country. The spectrum allocation dictates the system design and performance limit of a system.

Tab. 1.2: Main regulated spectrum allocation for broadcasting and communications.

Frequency span	Service/system
19.95–20.05 kHz 2,495–2,501 kHz 2,501–2,502 kHz 2,502–2,505 kHz 24.990–25.005 MHz 25.005–25.010 MHz	Standard frequency and time signals
535–1,606.5 kHz	AM radio
87–108 MHz	FM radio
54–88 MHz	Broadcast TV (Channels 2–6)
174–216 MHz	Broadcast TV (Channels 7–13)
450–470 MHz 698–862 MHz 790–862 MHz 3,400 MHz–3,600 MHz	IMT 3G and beyond 3G
1,920–2,170 MHz	3G Cell Phones (FDD)
1,880–1,920 MHz 2,300–2,400 MHz	3G Cell Phones (TDD)
470–678 MHz	5G Mobile (T-mobile, USA)
470–806 MHz	Broadcast TV (UHF)
746–764 MHz, 776–794 MHz	3G Broadband Wireless
1.7–1.85 MHz, 2.5–2.69 MHz	3G Broadband Wireless
885–905 MHz, 930–950 MHz	1G Analog Cellular Phones

Tab. 1.2 (continued)

Frequency span	Service/system
700 MHz	5G Mobile(European Union)
890–960 MHz	2G Digital Cellular Phones (GSM900)
880–890 MHz	2G Digital Cellular Phones (EGSM900)
1,710–1,880 MHz	2G Cell Phones (GSM1800)
2,320–2,325 MHz	Satellite Digital Radio
2,515–2,675 MHz	5G Mobile (China Mobile)
3,400–3,500 MHz	5G Mobile (European Union)
3,400–3,800 MHz	5G Mobile (China Telecommunications)
3,500–3,600 MHz	5G Mobile (China Unicom)
3,550–3,700 MHz	5G Mobile (NT&T, USA)
3,600–3,700 MHz	5G Mobile (NTT DoCoMo, Japan)
3,700–3,800 MHz	5G Mobile (KDDI, Japan)
3,800–3,900 MHz	5G Mobile (Rakuten, Japan)
3,900–4,000 MHz	5G Mobile (SoftBank, Japan)
4,000–4,100 MHz	5G Mobile (KDDI, Japan)
4,500–4,600 MHz	5G Mobile (NTT DoCoMo, Japan)
4,800–4,900 MHz	5G Mobile (China Mobile)
11.7–12.75 GHz	Digital Broadcast Satellite (Satellite TV)
12.75–13.25 GHz	Deep Space Communications
38.6–40 GHz	Fixed Wireless Services
40.5–42.5 GHz	Broadcast Satellite

1.2.2.2 Unregulated spectrum

Unlicensed spectral bands are allocated by the regulatory bodies also for one or several specific services (e.g., wireless local area network (WLAN)), but not a specific operator. Often different countries match their frequency allocation for the free band so that technology developed for that band is compatible but also not exactly the same. The main unlicensed spectrum bands include the ISM (industrial, scientific and medical equipment) band and the U-NII (the unlicensed national information infrastructure) band, listed in Tab. 1.3.

Since any device is allowed to operate in these bands without authority, this encourages technology innovation, speed-up system rollout and decrease the total cost. But the devices must limit their transmit power, follow rules for the signal restrictions in shape and bandwidth, and use the band according to the purposes stipulated by the regulatory bodies. This means that these systems must use an inefficient signaling scheme and have to suffer interference from other devices, thus may result in unfavorable performance.

Tab. 1.3: Unlicensed spectrum allocation.

Frequency span	Purpose/service	Remark
6.765–6.795 MHz 13.553–13.567 MHz 26.957–27.283 MHz 40.66–40.7 MHz	–	ISM band in Hong Kong and Macao, China
902–928 MHz	ISM band I (cordless phone, IG WLANs)	Part of this band is assigned to GSM system in Europe
2.4–2.4835 GHz	ISM band II (Bluetooth, 802.11b and 802.11g WLANs)	–
5.725–5.85 GHz	ISM band III (wireless PBX)	–
5.15–5.25 GHz	NII band I (Indoor systems, 802.11a WLANs)	–
5.25–5.35 GHz	NII band II (short outdoor and campus applications)	–
5.725–5.825 GHz	NII band III (long outdoor and point-to-point links)	–
5.725–5.850 GHz	–	ISM band in Hong Kong and Macao, China
5.850–5.875 GHz	–	ISM band in Macao, China
24–24.05 GHz 24.05–24.25 GHz 61–61.5 GHz 122–123 GHz 244–246 GHz	–	ISM band in Hong Kong and Macao, China

1.2.3 Energy constrains

One of the advantages of wireless communications is that there is no wire restriction between the transmitter and the receiver, but the mobile station needs to be powered by batteries to benefit from this advantage. This kind of power supply imposes requirements on the power consumption:

1) Power amplifiers with high power efficient, modulation formats insensitive to nonlinear distortions or related techniques should be considered. In the choosing power amplifiers, the power amplifiers with high power efficiency are preferred especially for mobile stations. However, the power amplifiers of efficiency higher than 50% are generally nonlinear, for example, Class-D, Class-E or Class-F amplifiers. As a consequence, the modulation formats adopted should be insensitive to nonlinear distortions. However, such kinds of modulation formats are generally spectrum inefficiency. If we

want to use modulation with high spectral efficiency in a high power efficiency mode, power amplifier linearization techniques such as predistortion, postdistortion, feedforward linearization, envelope elimination and restoration, and linear amplification with nonlinear components should be used.

2) The limitations of signal processing on energy efficiency and receiver sensitivity, as well as the requirements for high-quality batteries, have become important factors. To meet these requirements, low-speed electronic devices may be chosen. However, the advanced algorithms for interference suppression are limited. For the mobile receiver, it must have high sensitivity so that acceptable performance can be achieved in very low received power. This can lower the battery size, but can also impact the design of the receiver and the network. Fortunately, battery technology has also developed rapidly in recent years, which has been partly alleviated by the provision of large capacity, fast charging and wireless charging technology.
3) Adaptive power control policies should be adopted to lower the power consuming. The transmit power should be adapted to the channel state and requirement. For example, small power can be used when the transmitter and receiver are closer. For voice transmission, the device only transmit at the user actually talks; this is the so-called discontinuous voice transmission (DTX). For sensor networks, "standby" and "sleep" mode can be defined.

1.2.4 User mobility constrains

User mobility not only cause fading phenomenon but also bring other problems. One important consequence on the cellular system is the registration and handoff problem that the network needs to determine which cell a user is in and will be moved to.

1) The registration, roaming and handoff problem. In cellular system, there are two databases. The first one is the Home Location Register (HLR), and tt is a central database that keeps track of the location a user is currently at. The second one is the Visitor Location Register (VLR), and it is a database associated with a certain BS that notes all the users who are currently within the coverage area of this specific BS. For example, if a user registered in Xi'an but currently in Beijing, it will inform the nearby BS in the city of Beijing that the user is now in its coverage, and the BS will record this information in VLR and forward it to the HLR in Xi'an. If one makes a call, the system will first query its HLR to find out its current VLR. The call is then rerouted to its current location. This service is called roaming service. The handoff problem is similar. When a MS is at the range of one cell, assume it is moving away from the area covered by the cell and eventually entering the area covered by a second cell, the call should be transferred to the second cell in order that the call is not terminated when the MS gets outside the range of the first cell. This handoff should be performed without interrupting the call and be unnoticeable to

the user. This requires complicated signaling. The handoff problem also exists in the satellite mobile networks whereas the satellite who acts as the BS moves.

2) When the user moves at a very high speed of more than 250 km/h, such as the newly constructed high-speed rail in China, it not only causes big Doppler shift and needs complicated handoff techniques but also affects the design and planning of the network BS.

1.2.5 Cross-layer design

Perhaps the most significant technical challenge in wireless network design comes from the design process that is completely different from that of the wired networks. In wired network design, protocols associated with different layers of the system operation are designed separately. Each layer provides service interfaces to the upper layer. This makes design simple and standardized. However, the wireless channel is poor and complex, and the channel as well as the network topology is time varying. The designer should consider the channel and its variations adaptively. Thus, the design is generally integrating and adapting the protocols at all layers.

1.3 Wireless systems and standards

Currently, most people are subscribers of some kind of wireless communication systems. With old systems going into the history and new systems emerging, the design, cost, complexity, performance and types of service offered by these systems are vastly different. We will briefly discuss the most common systems in this section.

1.3.1 Cellular telephone systems

Cellular telephone systems are very popular worldwide and undoubtedly are the most successful and lucrative wireless systems. These systems provide two-way voice and data communications all over the world.

1.3.1.1 First-generation cellular systems

The important stage of cellular system is the World Allocation Radio Conference held in 1976. The 800/900 MHz band was approved to be allocated for the deployment of cellular systems at the conference. After then, the first-generation cellular systems based on frequency division multiple access (FDMA) and analog FM technology were deployed in many countries. NTT (Nippon Telegraph and Telephone Corporation) at Japan took the lead in 1979 by launching the first commercially automated cellular network (the 1G). In 1981, Nordic Mobile Telephone (NMT) system ranked the second 1G network which was launched in Denmark, Finland, Norway and Sweden. There were also several countries such as the UK, Mexico and Canada launched their 1G network in the early 1980s. China launched its 1G network in 1987. By using FDMA, there is one single channel per carrier in the 1G network.

The capacity of the first-generation system is greatly limited. The main consideration of 1G system construction is the cost of BS. That is the main reason that an entire city or region was covered by a fairly small number of cells in the cellular systems. The power of the cell BSs transmitted very high so that it covers an area of several square kilometers. The radiated signal power was uniformly districted in all directions so that all mobile stations at a circle around the BS have approximately same received power in case that the signal was not blocked by obstacles. Therefore, a hexagonal cell shape for the system which is close to a circle was produced by this circular contour of constant power, as shown in Fig. 1.4. The same channel, for example, frequency of the 1G network, can be reused in two cells that are certain distance apart.

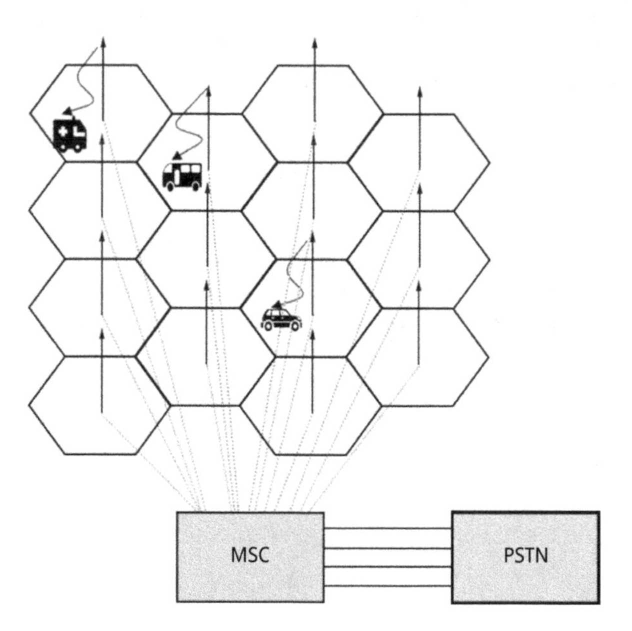

Fig. 1.4: Cellular systems.

1.3.1.2 Second-generation cellular systems

In the 1990s, the second-generation (2G) mobile phone systems emerged, and GSM standard (Global System for Mobile Communications) is the main 2G standard. Different from the 1G system, digital transmission and fast out-of-band phone-to-network signaling are used in these systems. In 2G period, the mobile phone usage was explosive and prepaid mobile phones appeared in this era. Since GSM was proposed in Europe, the first GSM network was launched by Radiolinja in Finland in 1991. Then, the GSM standard was adopted worldwide. The first GSM network in China was launched by China Mobile in 1994 in Guangdong province. The frequencies adopted by GSM were higher in Europe than those in America, although there was some overlap in phase 1 in which 900 MHz frequency band was used. The 900 MHz frequency range used in GSM was also used for the 1G system in Europe, so 1G systems were rapidly closed down to make space for the 2G systems. The second version of GSM is GSM1800 which used frequency band around 1800 MHz. The third version of GSM is GSM1900 which used frequency band around 1900 MHz was mainly used in America. There are also two other 2G standards: IS-54, which adopts a combination of TDMA and FDMA and phase-shift keying modulation, and IS-95, which adopts direct-sequence CDMA combined with binary modulation and coding.

With the introduction of 2G system, a major change was the "downsizing" of mobile phones, that is, from large "brick" mobile phones to small 100–200 g handheld devices. There were several reasons for this change. The first was due to technological progress, such as more advanced batteries and more energy-efficient electronic products. Secondly, smaller cells were adopted to meet the needs of increasing user capacity. A smaller cell means that the average transmission distance from the mobile phone to the BS was shortened, thus prolonging the battery life.

The 2G systems also provided a new form of communication called SMS or text messaging. It was first introduced in GSM networks but eventually spread to all digital mobile systems. Another contribution of 2G system is the introduction of the ability of accessing media content on mobile phones. In 1999, NTT DoCoMo in Japan first introduced full internet service on mobile phones. After that, high-rate packet data services have become standard services in all of the digital cellular standards. The data rates GSM systems provided can be up to 100 kbps, and this is implemented by aggregating all timeslots together for a single user. This enhancement method is named General Packet Radio Service (GPRS). There was another enhancement called Enhanced Data Services for GSM Evolution (EDGE) and it increases data rates by using a high-level modulation format combined with FEC channel coding. By a similar way, the IS-54 provided high data rates (HDRs) of 40–60 kbps by using both aggregation of time slots and high-level modulation. In the other 2G standard, the IS-95 systems used a time-division technique called HDR to support higher data.

1.3.1.3 Third-generation cellular systems and beyond

As the demand for greater data speeds increases, the 2G technology apparently is not qualified for the job, so the 3G technology is born. The 3G systems use packet switching for data transmission rather than the circuit switching which is main switching method in 2G systems. In addition, the 3G standardization process focused on the data rate requirements. For example, it was required to support 2 Mbps maximum data rate indoors and 384 kbps outdoors.

The research and development of 3G projects began in 1992. In 1999, five radio interfaces for IMT-2000 as a part of the ITU-R M.1457 Recommendation was approved, including WCDMA, CDMA 2000, TD-SCDMA, UWC-136 and DECT (Digital Enhanced Cordless Telecommunications). In 2007, WiMAX (Worldwide Interoperability for Microwave Access) was added as the sixth 3G standard. UWC-136 and DECT are only suitable for the upgrading of IS-136 and original DECT, and they do not fully satisfy the 3G specification of ITU. It can be seen that the standard process is a complex trade-off of interests between different countries or organizations. The WiMAX is actually a "last mile" solution for internet. WCDMA, CDMA 2000 and TD-SCDMA are the three main 3G cellular standards listed in Tab. 1.4.

Tab. 1.4: Overview of 3G/IMT-2000 standards.

Common name	**CDMA2000**	**WCDMA**	**TD-SCDMA**
ITU IMT-2000 standard	CDMA multicarrier (IMT MC)	CDMA direct spread (IMT-DS)	CDMA TDD (IMT TC)
Group	3GPP2	3GPP	
Bandwidth of data	Evolution-data optimized (EV-DO)	High-speed packet access (HSPA)	
Duplex	FDD (frequency-division duplexing)		TDD (time-division duplexing)
Mode of operation	Multicarrier and direct spreading DS-CDMA at a rate of aN = 1.2288 Mcps, with N = 1, 3, 6, 9 and 12.	DS-CDMA at a rate of N × 0.960 Mcps, N = 4, 8 and 16.	DS-CDMA at a rate of aN = 1.28 Mcps, with N = 1, 2, 4, 8 and 16.
Core network	Evolved ANSI-41	Evolved GSM MAP	
Key features	1. Backward compatibility with IS-95. 2. Downlink can be implemented using either multi-carrier or direct spreading. 3. Uplink can support a simultaneous combination of multi-carrier or direct spreading. 4. Auxiliary carriers to help with downlink channel estimation in forward link beamforming.	1. Wideband DS_CDMA system. 2. Backward compatibility with GSM/DCS-1900. 3. Up to 2.048 Mbps on Downlink. 4. Minimum forward channel bandwidth of 5 MHz. 5. Connection-dedicated pilot bits in downlink beamforming.	1. RF channel bit rate up to 2.227 Mbps. 2. Smart antenna technology is used (but not strictly required). 3. Pilot bits are not required since the channel reciprocity. 4. The uplink and downlink data rate can be adaptively changed for asymmetrical services. 5. Paired bandwidth allocations are not required; the bandwidth allocation is flexible. 6. No transceiver isolation request, RF can be implemented on a signal IC.
Geographical areas	Americas, Asia, some others	Worldwide	China, Romania, Korea, France, Italy

Although mobile phones had the ability to access data networks such as the Internet for a long time, they had not been widely popularized. It was not until the mid-2000s that specialized devices, such as the "dongles," plugged directly into a computer through the USB port, began to access the mobile Internet due to high-quality 3G coverage. Another new class of device, the so-called "compact wireless router," enables multiple computers to connect to the 3G Internet through WiFi at the same time rather than just connecting to one computer through a USB plug-in. Such devices have become particularly popular because laptops have increased portability. Therefore, some computer manufacturers began to embed mobile data functions directly into laptops and insert SIM cards directly into devices. Other types of data-aware devices, such as E-readers, tablet devices can also embed the wireless internet.

During the development of 3G cellular systems, 2.5G systems which were extensions to existing 2G networks were also developed, including CDMA2000 1× and GPRS. These 2.5G networks implemented partial functions of 3G networks but did not fully provide the promised HDRs or all the multimedia services. CDMA2000-1X supports a theoretical maximum data rate up to 307 kbps. The EDGE system meets the requirements of 3G system theoretically, but there is no practical system that has truly achieved these requirements. High-speed downlink packet access (HSDPA) which is an evolution of 3G technology, begun to be implemented in the mid-2000s. It is an enhanced version of 3G mobile cellular communications protocol in the high-speed packet access (HSPA) family, also called 3.5G, 3G+ or turbo 3G, which allows Universal Mobile Telecommunications System-based networks to support higher data rate and capacity. The HSDPA deployments provided down-link speeds of 1.8, 3.6, 7.2 and 14.0 Mbps. Higher data rate implementation are available with HSPA+, which supports data rates of up to 42 Mbps downlink and 84 Mbps with Release 9 of the 3GPP standards. Around the year of 2008, interest in broadband wireless internet access was sharply increased, and this led the development of fourth-generation cellular system: 3GPP Long-Term Evolution (LTE), LTE-Advanced and WiMAX.

In October 2010, two technology TD-LTE-Advanced and 802.16m were selected out from six candidates and approved by the ITU as the 4G technology standards. By the end of 2011, 4G International Standard Proposal Book was completed by ITU and was officially released at the beginning of 2012.

Currently, the fifth-generation mobile communication technology is the latest generation of cellular mobile communication technology. The performance goal of 5G is to increase data rate, reduce delay, save energy, reduce cost, increase system capacity and connect large-scale equipment. The first phase of the 5G specification in Release-15 was designed to accommodate early commercial deployments. The freezing of the second phase of Release-16 was announced on the evening of July 3, 2020. The freezing means the completion of R16 standard, and the 5G network has expanded from the connection between people to the connection between people and things, the connection between things and things, and the interconnection of all things has become possible. The data rate of 5G is up to 20 Gbps, which provides a technical basis for various new applications. Automatic driving, industrial Internet and other applications will be accelerated.

1.3.1.4 The call making process of a cellular network

Now, let's discuss the process of making and receiving a call. In Fig. 1.5, there are *mobile stations (MS), base stations (BS)* and a *mobile switch center (MSC).* The MSC is responsible for connecting all the MSs to the PSTN in a cellular network. Each MS communicates with a BS and maybe hand-off the others during a call. The MS generally consist of one transceiver and one antenna while the BS consists of several transmitters and receivers capable of handling full duplex communications and generally possess several transmitting and receiving antennas mounted on a tower. The BS connects all the mobile calls in the cell to the MSC. The MSC coordinates the activities of all of the BSs including the conversation, billing and system maintenance functions and connect the cellular system to the PSTN. In a large city, several MSCs may be used by a carrier. Four different channels specified by the standard *common air interface* defines the communication between the mobiles and the BS. They are listed in Tab. 1.5.

Tab. 1.5: Channels defined by the standard *common air interface.*

Channels	Abbreviation		Functions	Direction
Forward voice channels	FVCs		Voice transmission	BS to MS
Reverse voice channels	RVCs			MS to BS
Forward control channels	FCCs	Setup channels	Initializing the mobile calls/broadcast traffic request	BS to MS
Reverse control channels	RCCs		Initializing the mobile calls	MS to BS

When a cellular phone is turned on, but is not engaged in a call, it first scans the FCCs to determine the one with the strongest BS signal. It then monitors this channel and again scans the FCCs when the current FCC drops below a given level. Typically, there are 10–60 voice channels and one control channel in one cell's BS, and the FCCs in neighboring cells are different because of the reuse concept. Generally, the FCCs are defined and standardized throughout the entire geographic area, so every phone scans the same channels while idle. Figure 1.5 shows the process of how a call is initialized by a landline subscriber while Fig. 1.6 shows how a call is initialized by a mobile phone. In the figures, the mobile identification number is the telephone number to identify the subscriber, the ESN is the electronic serial number to identify the cellular terminal and the station class mark is to indicate the maximum power level for the particular user.

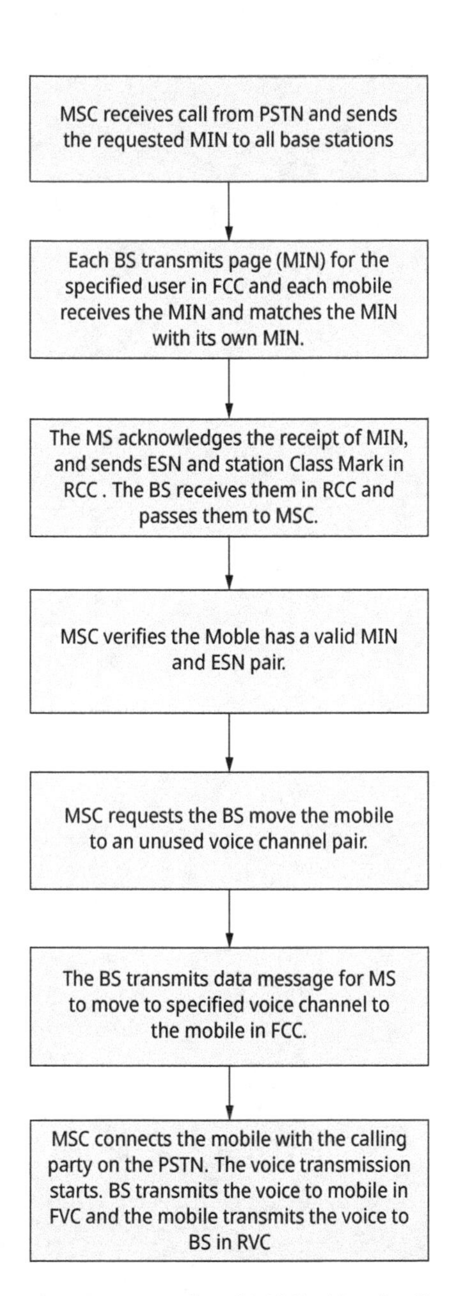

Fig. 1.5: Process of a call initialized by a landline.

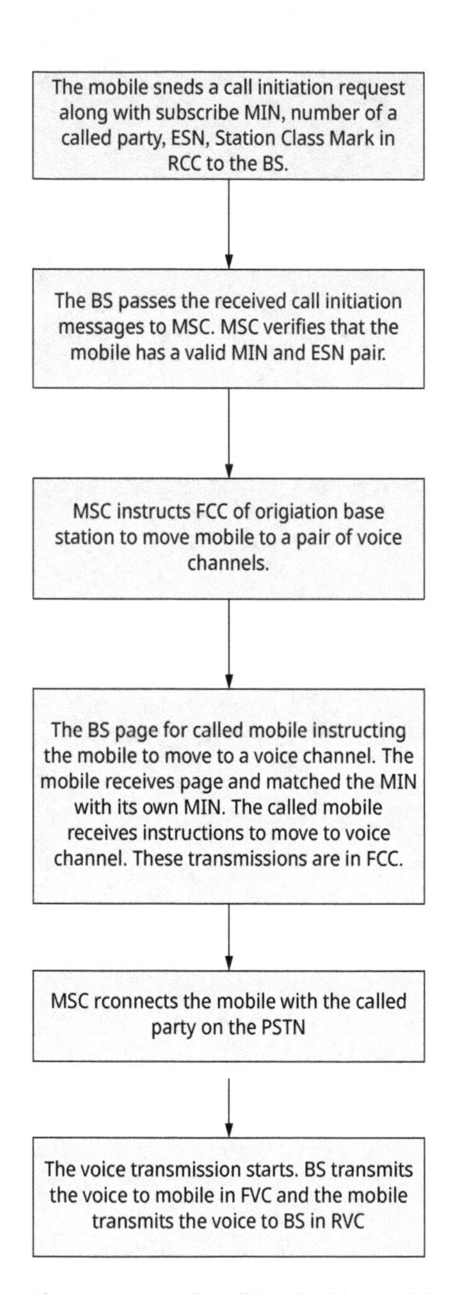

Fig. 1.6: Process of a call initialized by a mobile.

1.3.2 Wireless local area networks

Wireless LANs support high-speed data transmissions within a small region at hotpots through wireless connections. This enables users to move within the local coverage area and still be connected to the network.

In 1970, Professor Norman Abramson at the University of Hawaii developed ALOHA network which is the first packet radio network and the first wireless computer communication network. The first experimental WLAN was reported in 1979 in which the diffused infrared communications were used. In the next year, an experimental application using direct sequence spread spectrum (DSSS) for wireless terminal communications was reported. In 1985, FCC announced the commercial application of spread spectrum technology of the ISM bands. The first-generation WLANs were based on proprietary and incompatible protocols. As the secondary users operated on unlicensed bands, these WLANs used spread spectrum to keep their low power and cope with the interferences from other users.

1.3.2.1 WLAN architectures

Let's define some notations related to WLAN.

Stations

Stations are all components that can connect into a wireless medium in a network. A station is equipped with wireless network interface cards. Stations can be classified into two categories of stations: access points (AP) and clients.

Access points (AP)

APs can also be called routers, and they act as BSs for the wireless network. The wireless enabled devices can communicate with the wireless signals that APs transmit and receive.

Client

Clients can be any devices including mobile terminals such as intelligent phones, laptops, personal digital assistants or fixed devices such as desktops and workstations equipped with a wireless network interface.

Basic service set (BSS)

BSS is defined as a set of all stations that can communicate with each other. BSS can be classified into two categories: independent BSS (IBSS) and infrastructure BSS.

Each BSS has an identification, which is the MAC address of the AP servicing the BSS and is called BSSID.

Independent BSS (IBSS)

An IBSS is an ad hoc network without APs, which means that they cannot connect to any other basic service set.

Infrastructure BSS

An infrastructure BSS contains APs, through which they can communicate with other stations not in the same basic service set.

Based on the conceptions above, we can discuss the WLAN architectures consisting of the following three modes.

Peer-to-peer (P2P) **or ad hoc mode**

P2P denotes a decentralized mode whereby wireless devices can communicate with each other directly. Wireless devices can discover and communicate with each other directly without an intermediary central AP if they are in the range of each other. P2P is often used in two-computer scenario because they can connect to each other to form a network.

If all stations in a network communicate only P2P, the network is called ad hoc network. Fig. 1.7 shows a Ad-hoc mode WLAN. Since there is no centralized server, the communication is accomplished by using the IBSS without getting permission from a coordinating point.

Fig. 1.7: Ad-hoc mode WLAN.

Mesh mode

In mesh mode, every client in the network also acts as an access or relay point, creating a “self-healing” and at least in theory infinitely extensible network. Fig. 1.8 shows a mesh mod WLAN.

Infrastructure mode

This is the most popular form of WLANs, and Fig. 1.9 shows infrastructure mode WLAN. The AP acts as the hub of a “star topology.” Any communication has to go through AP. If a client, like a computer, or a phone, wants to communicate with another MS, it first needs to send the information to AP, then AP sends it to the destination MS. Multiple APs can be connected together and handle a large number of clients. More APs can work in a cellular mode to organize a larger network. Furthermore, the APs can be connected to internet through routers.

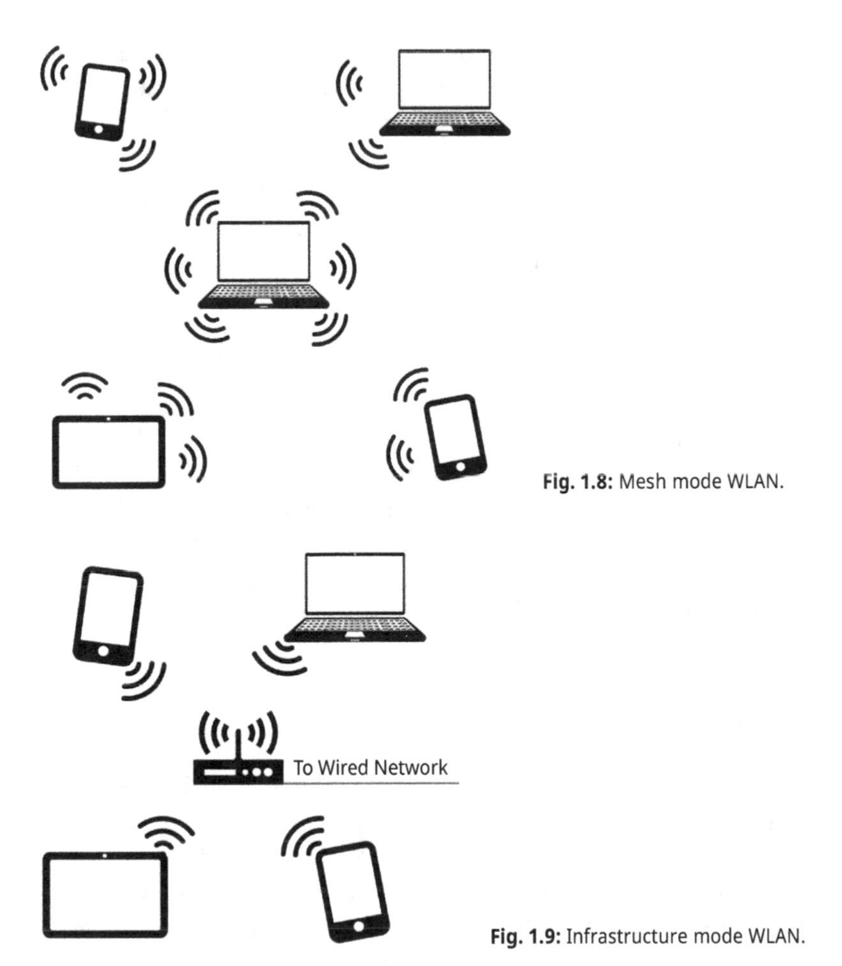

Fig. 1.8: Mesh mode WLAN.

Fig. 1.9: Infrastructure mode WLAN.

1.3.2.2 WLAN standards

Table 1.6 lists the WLAN standards. The standardization of the second-generation WLANs avoids some problems with the proprietary first-generation systems. From the table, we can find out the different techniques and different data rate transmissions that different standards adopt and support. Companies, universities and many hotspots such as tea houses, airport, hotels and even the bus have installed 802.11-based BSs to offer wireless access. The main cellular carrier also developed their WLANs to provide value-added services.

1.3.2.3 The multiple accesses in WLANs

As we know, WLANs are designed for data exchange; the quality of service (QoS) requirements are quite different from that of the voice conversations. In data communication, the data is grouped into packets/frames. Each packet/frame contains a number of bits of information. Before an MS (mobile station) sends its packets, it checks to see if someone else is sending information. Only when the medium is free can the MS send its packets. If some station is sending or receiving signal, the MS that intends to send will generate a random waiting time and wait for its turn. If several MSs are all waiting for their turns since their waiting times are randomly generated and thus not equal, they will not start sending simultaneously. Thus collision (two or more MSs sending signals simultaneously) is avoided. That's why it's called carrier sensing multiple access with collision avoidance (CSMA/CA).

When two MSs cannot hear each other (e.g., blocked by a wall), by exchanging request to send (RTS)/clear to send (CTS) mechanism, CSMA/CA can optionally be supplemented to avoid collision. A terminal ready for transmission sends an RTS packet identifying the source address, destination address and the length of the data to be sent. The destination station responds with CTS packet. The source terminal receives the CTS and sends the data. Other terminals go to the virtual carrier-sensing mode (NAV signal on); therefore, the source terminal sends its packet with no contention. After completion of the transmission, the destination station sends an ACK, opening contention for other users. The process is shown in Fig. 1.10.

1.3.3 Codeless phones

A cordless telephone is an extension to fixed phone. The BS attaches to the telephone network exactly the same way as a corded telephone does and is on the subscriber premises. One or more wireless handsets communicate with a BS via radio waves when they are within a limited range of its BS (which has the handset cradle).

The main difference between a cordless telephone and a mobile telephone lies in the BS on subscriber premises. However, the difference between cordless and mobile telephones had been made burred by the main cordless telephone standards, such as Personal Handy-phone System (PHS) and DECT, in which cell handover and various advanced features, such as data-transfer and even, international roaming on a limited scale were implemented. In these models, a commercial mobile network operator is responsible for maintaining the BSs and users subscribe to the service. Do not like the corded telephone, it is necessary for a cordless telephone to main electricity for powering the BS. The cordless handset is powered by a rechargeable battery, and it is charged when sitting in its cradle.

Cordless phone was invented in 1965 by Teri Pall who was a jazz musician, but the invention caused radio signals to interfere with aircraft for its two-mile range. The other two inventors, an amateur radio operator George Sweigert, and an inventor from Cleveland, Ohio, are largely recognized as the father of the cordless phone for its patent "full duplex wireless communications apparatus" in 1969. In the first cordless systems, one phone handset can connect to each base unit with coverage of several rooms of a house or office. Since there is no coordination within cordless phone

Tab. 1.6: WLAN standards.

Standards	802.11	802.11b	802.11a	802.11g	802.11n	
Released time	July 1997	Sept 1999	Sept 1999	June 2003	Oct 2009	
Legal bandwidth	83.5 MHz	83.5 MHz	325 MHz	83.5 MHz	20 MHz	40 MHz
Frequency range	2.400–2.483 GHz	2.400–2.483 GHz	5.150–5.350 GHz 5.725–5.850 GHz	2.400–2.483 GHz	5 GHz or 2.4G mode	
Number of nonoverlapping channels	3	3	12	3	2	
Modulation techniques	Frequency hopping spread spectrum/direct sequence spread spectrum (FHSS/DSSS)	Complementary code keying/ direct sequence spread spectrum (CCK/DSSS)	Orthogonal frequency division multiplexing (OFDM)	CCK/OFDM/ DSSS	(BPSK, QPSK, 16QAM, 64QAM)-MIMO-OFDM	
Physically transmitting rate (Mbps)	1, 2	1, 2, 5.5, 11	6, 9, 12, 18, 24, 36, 48, 54	6, 9, 12, 18, 24, 36, 48, 54	7.2, 14.4, 21.7, 28.9, 43.3, 57.8, 65, 72.2 per stream, up to four streams in total	15, 30, 45, 60, 90, 120, 135, 150 per stream, up to 4 streams in total

(continued)

Tab. 1.6 (continued)

Standards	**802.11**	**802.11b**	**802.11a**	**802.11g**	**802.11n**
Antenna coverage range	20 m indoor 100 m outdoor	38 m indoor 140 m outdoor	35 indoor 120 outdoor	38 m indoor 140 m outdoor	70 m indoor 250 m outdoor
Maximum UDP throughout	1.7 Mbps	7.1 Mbps	30.9 Mbps	30.9 Mbps	–
Maximum TCP/IP throughout	1.6 Mbps	5.9 Mbps	24.4 Mbps	24.4 Mbps	–
Compatibility	N/A	Compatible with 802.11g	Incompatible with 802.11b/g	Compatible with 802.11b	Compatible with 802.11a/b/g

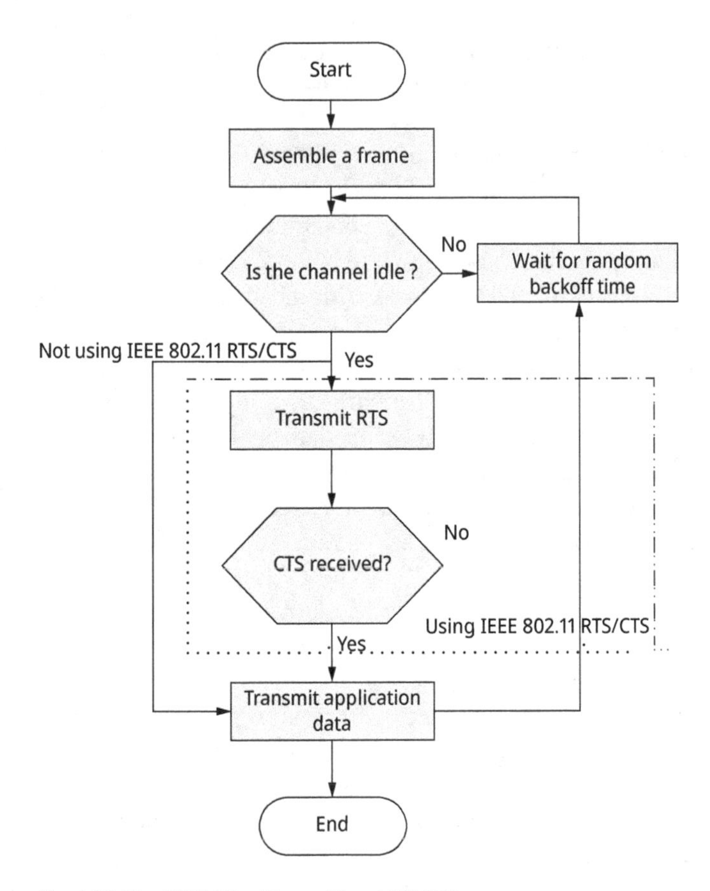

Fig. 1.10: The CSMA/CA with or without RTS/CTS.

systems, a high density of these systems will cause interference between them. This is why later cordless phones were designed to have multiple voice channels and scan between these channels to select the channel with the least interference. They also use spread spectrum technology to reduce interference from other systems.

In the early 1990s, the second-generation cordless phone standard CT-2 was developed in some countries in Europe providing short-range proto-mobile phone service. Typical CT-2 users bought a handset and a telepoint (may also be called BS), and they connected the telepoint to their own home telephone system. Cordless phone would work exactly the same as a standard cordless phone, where the calls from the home phone point were routed via the home telephone line. When the user is in the coverage of the telepoint, the user could receive incoming calls. If it is out of range of the home, the CT-2 user was capable of finding signs indicating a network BS in the area, generally in hot points such as shopping malls, busy streets, gas stations, train stations and airports and make outgoing calls (but not receive calls) using the network BS. In this way, the cordless phone handset can be used at home and away from home to make calls with a lower rate than that of cellular service. CT-2 is a digital FDMA system that uses Time Division Duplexing technology to share carrier frequencies between handsets and telepoints.

European DECT system is another evolution of the cordless telephone, which is primarily designed for office buildings. The DECT standard gives a full specification to allow a cordless telephone to access a fixed telecoms network through wireless. Different from the standards of the cellular mobile phone system, it does not provide specifications for internal aspects of the fixed network. This means that there are different options for connecting to a fixed network, for example, it can be through a BS, "Radio Fixed Part," or a gateway. Gateway connections might have different options, such as PSTNs, telephone jacks, or voice over IP. There are other applications utilizing DECT such as baby monitors, and there is no gateway functionality in these devices.

In some countries such as India and South Africa, DECT has been used as a substitute for copper pairs in the "last mile" for fixed wireless access. The cell coverage might be extended to over 10 km by using directional antennas and at the same time sacrificing some capacity.

PHS is another advanced cordless telephone system invented and first deployed in Japan at July 1995. It is also marketed as the Personal Access System and commercially branded as Xiaolingtong in China.

PHS is essentially a cordless telephone like DECT. However, it provides handover the capability to support the user move from one cell to another during a connection. The maximum transmission power of PHS BS is as small as 500 mW so that the cells are small. The cell sizes typically range around tens or at most hundreds of meters, and some might range up to about 2 km in LOS. Note that CDMA and GSM cells typically range multikilometers. This feature makes PHS a suitable communication method for dense urban areas, but an impractical method for rural areas. Moreover, the small cell size makes it difficult to make calls from rapidly moving vehicles.

The radio channel access method adopted in PHS is TDMA/TDD, and voice codec used in PHS is 32 kbps ADPCM. The modern PHS phone also support many value-added services like e-mailing, text messaging, WWW access, color image transfer and other high speed wireless data/Internet connection (64 kbps and higher).

PHS technology is also used for bridging the "last mile" gap between the POTS network and the subscriber's home, which is local loop. It was also developed for providing a wireless front-end of an ISDN network. In this way, a BS of PHS is compatible with ISDN and is often connected directly to ISDN telephone exchange equipment, for example, a digital switch.

PHS is a microcellular system with low-cost BS which adopts "Dynamic Cell Assignment." Different from the typical cellular telephone systems, PHS provides higher number-of-digits frequency with efficiency of lower cost based on per area criterion. In Japan, it offers the flat-rate wireless service like AIR-EDGE.

The data rate of AIR-EDGE is accelerated by combining lines, each of which is basically 32 kbps. The first version of AIR-EDGE introduced in 2001 provides only 32 kbps service. In the next year, 128 kbps service (AIR-EDGE 4×) was provided. In 2005, 256 kbps (AIR-EDGE 8×) service was provided.

In 2006, the speed of each line was 1.6 times of the original version with the "W-OAM" technology. With the latest "W-OAM" capable instrument, the speed of AIR-EDGE 8× is up to 402 kbps.

"W-OAM typeG" was introduced in April 2007, which enables data rate of 512 kbps for AIR-EDGE 8× users. Moreover, the maximum data rate of the "W-OAM typeG" AIR-EDGE 8× service can be up to 800 kbps, and this enables higher data rate for APs to support switching lines from ISDN to fiber optic. In this scenario, the data rate may exceed the speed of popular W-CDMA 3G service like NTT DoCoMo's FOMA in Japan.

1.3.4 Satellite networks

Satellite communications are another important form of wireless communications. Modern satellite communications adopt a variety of orbits including GEO, MEO, LEO and other elliptical orbits.

Actually, the concept of the geostationary communications satellite was first proposed in October 1945 by a science fiction author named Arthur C. Clarke in his article entitled "Extra-Terrestrial Relays" for Wireless World. On August 19, 1964, the Syncom 3 was launched as the first truly geostationary satellite. Its orbit was in at 180° east longitudes over the international date line. The experimental television coverage of the 1964 Summer Olympics held in Tokyo, Japan was relayed to the United States so that these Olympic Games were the first to be broadcast internationally. Syncom 3 is sometimes recognized as the first television transmission to cross the Pacific Ocean; however, the first broadcast from the United States to Japan is the Relay 1 satellite which was done on November 22, 1963. On April 6, 1965, the first geostationary satellite for telecommunications over the Atlantic Ocean, Intelsat I, also known as Early Bird, was launched and placed in orbit at 28° west longitudes. On May 30, 1974, the first geostationary communications satellite in the world, the experimental satellite ATS-6 built for NASA, was launched, which was three-axis stabilized. The current GEO satellite systems include Immarst and OmniTRACS.

It is difficult to provide voice and data services over geosynchronous satellites for several main reasons. The two main drawbacks include the big round-trip propagation delay, large and bulky handset for high power and low data rates. So far, only a very few mobile phones can provide synchronous orbit satellite voice services for emergency communication, such as the Huawei Mate 60 Pro phone released on August 29, 2023, which is the first smartphone to provide synchronous satellite direct voice communication. For these reasons LEO satellites were developed to match the requirements of delay, power and data rates of voice and data communications. LEO satellites deliver significant voice quality over the GEO satellite systems. Table 1.7 lists the main characteristics of LEO communication networks in different applications. Among them, Globalstar and Iridium systems attracted the most attention.

The Globalstar Satellite constellation orbits 1,410 km above the earth's surface. It takes less than 2 h for a Globalstar satellite to complete a full rotation. Their relative low orbit indicates small propagation delay and path loss, which are suitable for the highest voice and quality available in the industry. Even using low-powered handset a user can enjoy high-quality voice clarity similar to a digital cellular phone.

The Iridium Satellite system attracted much attention when it was set up. Its constellation orbits 780 km above the earth's surface. An Iridium satellite travels at 17,000 miles an hour and it will orbit from pole to pole in 100 min. The Iridium constellation communicates with each other by using intersatellite links, which is different from that of the Globalstar constellation. Iridium offers the total planetary coverage because of the intersatellite links. Each satellite had four intersatellite links, two are used to communicate with satellites on either side and two are used to communicate to other satellites fore and aft in the same orbital plane. The disadvantage of intersatellite links is if the signal links through numerous satellites before transmission to an Iridium ground station the user may experience echo.

The Globalstar constellation consists of 48 LEO satellites that orbits have an inclination of 52 ° compared with Iridium's near-polar 86.4 ° orbits. Globalstar does not cover the poles due to the lower orbital inclination. The Globalstar satellites have no intersatellite linking like the Iridium satellites have. The Globalstar satellites are simply bent pipe repeaters. The Iridium constellation consists of 66 active satellites in LEO orbit. Given the Globalstar and Iridium satellites speed and the vast number of satellites in orbit, any coverage gaps are rare and are corrected in minutes.

Another very hot direction is to provide internet services by utilizing a huge number of low-orbit satellites. The typical example is the Starlink plan proposed by SpaceX (Space Exploration Technologies Inc.), which aims to launch around 12,000 satellites to form a "Starlink network" between 2019 and 2024 to serve the Internet. However, as of 11: 36 on August 17, 2023, Beijing time, SpaceX has launched 99 batches of Starlink satellites, with a total of 4,962. The orbital heights of these satellites are between 500 and 1,200 km, and these satellites and ground BSs will be interconnected to form a huge communication network for seamless communication coverage.

1.3.5 Short-range wireless communication technology: Bluetooth, ZigBee, RFID and Ultra-Wideband Radios

In the general sense, as long as both sender and receiver communications through the radio waves to transmit information, and the transmission distance is limited to short range, typically tens of meters or less, it can be called short-range wireless communications (Kraemer et al. 2009). If their cost and power consumption are decreased, it would be feasible to embed them in more types of electronic devices, which can be used to create smart homes, sensor networks and other compelling applications. We will discuss several of these technologies including Bluetooth, ZigBee, radio-frequency identification (RFID) and Ultra-Wideband Radios.

1.3.5.1 Bluetooth

Bluetooth is an open wireless standard, which can realize short-distance data exchange between fixed equipment, mobile equipment and building personal area network. Bluetooth can connect multiple devices, which overcomes the problem of data synchronization. This technology was originally created by telecoms vendor Ericsson in 1994 as a wireless alternative to RS-232 data cables.

Frequency-hopping spread spectrum is used in Bluetooth operating at the 2.4 GHz ISM band. Originally, Gaussian frequency-shift keying format was employed as the only modulation scheme. Then, $\pi/4$-DQPSK and 8DPSK modulation may also be used between compatible devices in Bluetooth 2.0+ EDR. In Bluetooth standard, the packet-based protocol with a master-slave structure was adopted. One master may communicate with up to seven slaves in a piconet. In the communication, the master's clock is shared by all devices. The synchronization of the Packet exchange is based on the basic clock defined by the master, which ticks at 312.5 μs intervals. In the simple case of single-slot packets, in even slots the master transmits and the slave receives; in odd slots the salve transmits and the master receives. Packets may be 1, 3 or 5 slots long but in all cases the master transmission will begin in even slots and the slave transmission begins in odd slots.

1.3.5.2 ZigBee

ZigBee is another low-cost, low-power standard based on the IEEE 802.15.2, and it is specifically designed for wireless mesh networking. It is designed for RF applications that require low data rates, long battery life and secure networks. ZigBee also operates in ISM bands:868 MHz in Europe, 915 MHz in the USA and Australia and 2.4 GHz in most places worldwide. People proposed Zigbee for simpler and less expensive technology than Bluetooth. It provides communications with data rates of up to 250 kbps within the range of 30 m.

1.3.5.3 Radio-frequency identification (RFID)

RFID is another short-range wireless technology intended to exchange data between a reader and an electronic tag attached to an object by using radio waves for the purpose of identification and tracking. The reader can read the tags from several meters away and even beyond the LOS. Parallel reading of tags is enabled by the application of bulk reading. Radio-frequency identification involves readers (or namely interrogators) and tags (or namely labels). There are at least two parts in most RFID tags. The integrated circuit part is designed for storing and processing information, modulating and demodulating an RF signal and other specialized functions. The antenna part is for receiving and transmitting the signal. RFID is becoming increasingly popular for the low cost. It has found wide applications in the payment by mobile phones, transportation payments, car-sharing, season parking tickets, product tracking, inventory systems, etc.

1.3.5.4 Ultra-wideband (UWB)

Ultra-wideband (UWB) is also a short-range wireless communication technology designed for transmitting information with a large bandwidth greater than 500 MHz or 20% of the center frequency. There are two forms of UWB implementations, one is called pulse-based UWB – wherein each transmitted pulse instantaneously occupies the UWB bandwidth, and the other is orthogonal frequency-division multiplexing (OFDM)-based UWB – an aggregation of at least 500 MHz worth of narrow band carriers. Unlike carrier-based systems that may suffer both deep fades and ISI, pulse-based UWB systems benefit from relative immunity to multipath fading (but not to ISI) because each pulse occupies the entire UWB bandwidth.

UWB was first proposed for personal area networks and this use appeared in the IEEE 802.15.3a draft PAN standard. Unfortunately, the IEEE 802.15.3a task group was dissolved in 2006. This work was transferred to the WI Media Alliance and the USB Implementer Forum. The development of USB-based consumer products is highly limited because the progress in UWB standards is low, initial implementations has a high cost and the performance is significantly lower than initially expected so that several UWB vendors ceased operations during 2008 and 2009.

Tab. 1.7: LEO communication networks.

System catalog	Little LEOs for nonvoice services			Big LEOs for voice plus limited data services			Broadband LEOs for high-speed data plus voice	
System	Orbcomm	LEO One	Final analysis	Iridium	Globalstar	Constellation communications	SkyBridge	Teledesic
Company	Orbcomm, Orbital Sciences	LEO One	Final analysis, general Dynamics info systems	Motorola	Loral, Alcatel, Qualcomm	Orbital Sciences, Bell Atlantic, Raytheon	Alcatel, Loral	Motorola, Boeing, Matra Marconi, Gates, McCaw
Service types	messaging, paging, e-mail	messaging, paging, e-mail	messaging, e-mail, file transfer	voice, data, fax, paging, messaging	voice, data, fax, paging, messaging	voice, data, fax	internet access, voice, data, video, videoconferencing	internet access, voice, data, video, videoconferencing
Voice (kbps)	–	–	–	2.4	adaptive 2.4/4.8/9.6	2.4	TBD	16
Data rate (kbps)	2.4 kbps uplink 4.8 kbps downlink	2.4–9.6 kbps uplink 24 kbps downlink	TBD	2.4	7.2	28.8	2,000 uplink 20,000 downlink	2,000 uplink 64,000 downlink
Orbit altitude (km)	825	950	1,000	780	1,410	2,000	1,469	1,375
No. of satellites	48	48	38	66	48	46	80	288

(continued)

Tab. 1.7 (continued)

System catalog	Little LEOs for nonvoice services			Big LEOs for voice plus limited data services			Broadband LEOs for high-speed data plus voice	
No. of orbit planets	3	8	7	6	8	8	2	12
Beams/ satellite	1	1	1	48	16	24/32	~50	64
Mobile uplink	148–150 MHz	148–150 MHz	VHF/UHF	1,616–1,626.5 (L-band)	1,610–1,626.5 (L-band)	2,483.5–2,500 (S-band)	–	–
Mobile downlink	137–138 MHz, 400 MHz	137–138 MHz	137–138 MHz	1,616–1,626.5 (L-band)	2,483.5–2,500.0 (S-band)	1,610–1,626.5 (L-band)	–	–
Feeder uplink	148–150 MHz	148–150.5 MHz	VHF/UHF	27.5–30.0 (Ka-band)	5.091–5.250 (C-band)	5.091–5.250 (C-band)	Ku-band	28,600–29,100 (Ka-band)
Feeder downlink	137–138 MHz, 400 MHz	400.15–401 MHz	VHF/UHF	18.8–20.2 (Ka-band)	6.875–7.055 (C-band)	6.924–7.075 (C-band)	Ku-band	18,800–19,300 (Ka-band)
ISL	No	No	No	Yes (4)	No	No	No	Yes (8)
Service date	1996	2002	2001	1998	2000	2001	2001	2004

Problems

1.1 Read material on paging systems and answer the questions:
   (a) Why do paging system need to provide low data rates?
   (b) How does a low data rate lead to a better coverage?
   (c) Why are small RF bandwidths used in transmitting messages in the paging systems?
1.2 What is the necessity of having a large number of cell-sites in a cellular system?
1.3 Describe disadvantages of using a wireless LAN instead of a wired LAN. And state the reasons that why the first-generation WLAN is not widely used.
1.4 (a) Summarize the key differences between the first-generation analog cellular and second-generation digital cellular system.
   (b) What are the core technologies for 5G systems to achieve high speed? What new applications might it have?
1.5 Explain the conditions for satellites to become geostationary orbit satellites. Is it convenient for voice applications and why?
1.6 In your place of study, how many modern wireless communications networks are available to you? Identify the type of services, the types of technologies, the commercial names of the service providers and the commercial names of the equipment manufactures that offer these wireless access capabilities.
1.7 Create a table list all 2G, 2.5G, 3G, 4G and 5G mobile standards. Carefully research the latest material to determine the parameters of each standard: (1) RF bandwidth and data rate; (2) modulation and coding types; (3) duplexing types; (4) new technologies from the previous generation; (5) main advantages and disadvantages.

Chapter 2
Wireless propagation channel modeling: large-scale path loss and shadowing

In high-speed wireless communications, the wireless radio channel is the fundamental limitation on the performance. The transmission between the transmitter and the receiver can be susceptible to noise, interference and other obstructions from buildings, mountains, foliage and so on. Because of the user movement as well as the environment dynamics, the changes caused by these channel impediments to channel over time are generally unpredictable. In this chapter, we will discuss the path loss and shadowing effects that can cause power variation in the received signal over the distance. Path loss is due to the dissipation of the power radiated by the transmitter and the effects of the channel propagation. Shadowing is due to obstacles between the transmitter and receiver that attenuate signal power through the effect of absorption, reflection, scattering, and diffraction. The power variation caused by path loss and shadowing occurs over large distance; however, the former varies in the distance of 100–10,000 m, and the latter is proportional to the size of the obstacles (10–100 m in outdoor environments and less in indoor environments). Since these two kinds of variations occur over relatively large distance, they are generally referred as *large-scale propagation effects*.

2.1 Radio wave propagation and frequency band for radio communications

2.1.1 Introduction to radio wave propagation

Theoretically, the radio wave propagation characteristics can be obtained by solving the famous Maxwell's equation with boundary conditions that express the physical details of the obstruction objects. Unfortunately, the information necessary for calculating the radar cross section of large and complex structures is difficult to get, and thus makes this method almost impossible. The main mechanisms that cause the electromagnetic wave propagation are reflection, diffraction, and scattering. Ray tracing approximation has been developed as the most common alternative method by representing the wave fronts as simple particles to determine the reflection and diffraction effects but generally ignore the more complex scattering phenomenon. To capture the more channel features in more complex environments, analytical models based on empirical measurements were developed. Shadowing caused by obstructions is too difficult to be described by deterministic channel model, so statistical models are often used. For the small-scale fading caused by constructive and destructive interfer-

https://doi.org/10.1515/9783110751437-002

ence by the large number of multipath components, statistical models are also used, which will be treated in the next chapter.

2.1.2 Frequency band for radio communications

The frequency band is a main consideration for predicting the effectiveness of radio communication links that we consider. The optimal frequency band for each propagation channel is determined and limited by the technical requirements of each communication system and by the conditions of radio propagation through each channel (Jakes 1974, Parsons 1992, Stüber 2002). Table 2.1 lists the spectrum of radio frequencies and their practical use in various communication channels. In this book, the models considered mainly for signals reside in the UHF (from 300 to 3000 MHz) and SHF band (from 3 to 30 GHz). The first band is favorable for terrestrial systems for it is not affected by the earth's curvature, and the second band is favorable for satellite applications for its ability to penetrate the ionosphere. Besides the propagation characteristics, antenna size is another determinant consideration (Blaunstein et al. 2007). The antenna size is constrained by signal frequency, since the antenna size for good reception is proportional to the signal wavelength. Therefore, moving systems to higher frequency band will allow for smaller antennas. On other hand, the received signal power in nondirectional antenna scenario is inversely proportional to the square of the frequency, so it is difficult to cover large ranges with higher frequency signals.

Tab. 2.1: The spectrum of radio frequencies and their practical use.

Frequency band	Frequency span	Characteristics and practical uses	Notes
Extremely low frequency (ELF)	<3 kHz	The wave propagates through the wave guide formed by the Earth's surface. and the ionosphere at long distances with a low degree of attenuation (0.1–0.5 per 1000 km.	–
Very low frequency (VLF)	3–30 kHz		–
Low frequencies (LF)	30–3000 kHz	In the 1950s and 1960s they were used for radio communication with ships and aircraft, but since then they are used mainly with broadcasting stations.	Because such radio waves propagate along the ground surface, they are called "surface" waves.

Tab. 2.1 (continued)

Frequency band	Frequency span	Characteristics and practical uses	Notes
High frequencies (HF)	3–30 MHz	Signals propagate by means of reflections caused by the ionospheric layers and are used for communication with aircraft and satellites, and for long-distance land communication using broadcasting stations.	–
Very high frequencies (VHF)	30–300 MHz	This frequency band is used for TV communication, in long-range radar systems and radio navigation systems.	–
Ultra high frequencies (UHF)	300–3000 MHz	This frequency band is very effective for wireless microwave links, constructions of cellular systems (fixed and mobile), mobile–satellite communication channels, medium range radars, and other applications.	–
C, X, K-bands (Ka, Ku).	3 GHz to several hundred gigahertz	These bands are used to construct modern wireless communication, such as satellite communications, inter-satellite links.	microwaves

2.1.3 Doppler frequency shift

When the transmitter or receiver is moving, the relative motion between them may cause a Doppler shift of $f_d = v\cos\theta/\lambda$ on the received signal, where θ denotes the arrival angle of the received signal relative to the direction of motion, v is the receiver velocity toward the transmitter in the direction of motion, and $\lambda = c/f$ is the signal wavelength ($c = 3\times10^8$ m/s is the speed of light, f is the transmitted signal frequency). Figure 2.1 shows the principle. Suppose in a short time interval Δt, a slight distance change in $\Delta l = v\Delta t\cos\theta$ that the transmitted signal needs to travel to the receiver will be caused by the movement of the receiver antenna mounted on the top of the car. A phase change of the resultant received signal introduced by the difference in path lengths is therefore

$$\Delta\varphi = (2\pi/\lambda)\times\Delta l = (2\pi/\lambda)v\Delta t\cos\theta = 2\pi v\Delta t\cos\theta/\lambda \tag{2.1}$$

The Doppler shift, or the change in frequency radiated, is given by f_D,

$$f_D = \frac{\Delta\varphi}{2\pi\Delta t} = v\cos\theta/\lambda \tag{2.2}$$

It is important to note that in Fig. 2.1 the angle $\theta = \theta'$ which means the angle keeps unchanged when the car moves from point X to point Y, and this is only being true when transmitter is far away from the moving antenna at points X and Y. The Doppler shift can be positive or negative depending on whether the mobile receiver is moving toward or away from the transmitter.

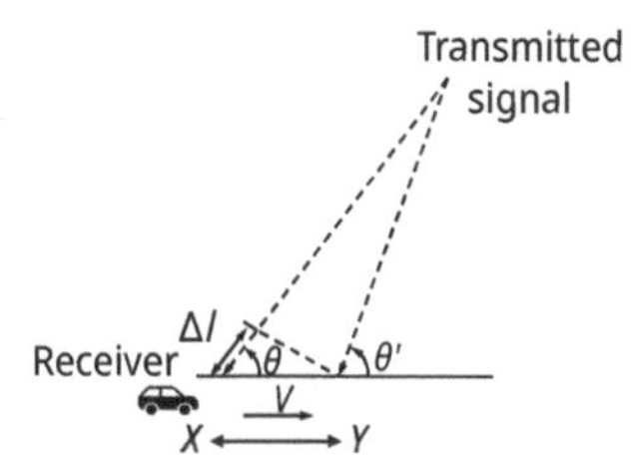

Fig. 2.1: Geometry-associated Doppler effect of the mobile link.

2.1.4 Equivalent low-pass representation of passband signals

In this book, we consider all the transmitter and received signals are passband real signals which do not include the baseband and UWB signals. This is true for most modulated signals because they are generated by using real sinusoid signal generating oscillators. Then we define the real modulated and demodulated signals are the real part of a complex equivalent baseband signal unconverted to the carrier frequency.

Let $s(t)$ be the transmitted signal, and it can be written as

$$\begin{aligned} s(t) &= \mathrm{Re}\left\{u(t)\exp^{(j2\pi f_c t)}\right\} \\ &= \mathrm{Re}\{u(t)\}\cos(2\pi f_c t) - \mathrm{Im}\{u(t)\}\sin(2\pi f_c t) \\ &= s_I(t)\cos(2\pi f_c t) - s_Q(t)\sin(2\pi f_c t) \end{aligned} \tag{2.3}$$

where $u(t) = s_I(t) + js_Q(t)$ is the equivalent low-pass representation of the passband signal $s(t)$, with bandwidth B_u equal to the bandwidth of $s(t)$ and power P_u is two times of the power of $s(t)$ P_t.

2.2 Free space propagation model

Consider a transmitter (TX) and a receiver (RX) are located in the free space with a distance d and there are no obstacles between them so that there is only a line-of-sight (LOS) ray from the transmitter to the receiver. We call the received signal LOS signal or ray.

According to the energy conservation law, we know that the power density integral over any closed surface surrounding the transmitting antenna should be constant and must be equal to the transmitting power.

Let's assume there is a sphere of radius d, and the center is the TX antenna. The TX antenna is supposed to be omnidirectional, and the power density on the surface is $P_{TX}/(4\pi d^2)$. The "effective area" of the RX is A_{RX}, and the power of the signal radiated to this effective area is collected by the RX antenna. Then the received power is then

$$P_{RX} = P_{TX}\frac{1}{4\pi d^2}A_{RX} \tag{2.4}$$

In case the transmit antenna is not isotropic, an antenna gains G_{TX} in the direction of the receive antenna should be multiplied:

$$P_{RX} = P_{TX}G_{TX}\frac{1}{4\pi d^2}A_{RX} \tag{2.5}$$

Effective isotropically radiated power (EIRP) or equivalent isotropically radiated is often used which is defined as the product of transmit power and transmit antenna gain in the considered direction. Another parameter, effective radiated power (ERP), is similar, but it is the radiated power as compared to a half-wave dipole antenna, instead of an isotropic antenna which is used in EIRP. As a dipole antenna has a 2.15 dB gain over an isotropic one, the value of ERP is 2.15 dB smaller than that of EIRP for the same system. In practice, we use antenna gains in units of dBi (dB gain respect to an isotropic antenna) or dBd (dB gain with respect to a half-wave dipole).

The relationship between the effective antenna area and the antenna gain can be found in Stutzman et al. (1997):

$$G_{RX} = \frac{4\pi}{\lambda^2}A_{RX} \tag{2.6}$$

Substituting eq. (2.6) into (2.5) yields

$$P_{RX} = P_{TX}G_{TX}G_{RX}\left(\frac{\lambda}{4\pi d}\right)^2 \tag{2.7}$$

This equation is the free space path loss model, which shows that the received power P_{RX} is a function of the distance d. It is also known as Fries' law.

It can be seen that the received power falls off proportional to d^2, where d is the distance between the transmit and receive antenna. We will see in the following sections that the received power falls much fast for other path loss models. The received signal power is also proportional to the square of the signal wavelength, or inversely proportional to the carrier frequency. This is easy to understand from eqs. (2.5) and

(2.4). Assume the antenna gain is given, and if we want to keep the same antenna gain, the effective area will be bigger for a larger wavelength. Of course, directional antennas can be designed to make the received signal power as an increasing function of frequency.

Let's define several concepts relative to path loss. Firstly, the *linear path loss* of channel as of transmit power to receive power is

$$P_L = \frac{P_{\mathrm{IX}}}{P_{\mathrm{RX}}} \tag{2.8}$$

The *path loss* in decibels is

$$P_L(\mathrm{dB}) = 10\log_{10}\frac{P_{\mathrm{TX}}}{P_{\mathrm{RX}}} \tag{2.9}$$

The path gain is defined as the negative of the dB path loss, which is

$$P_G(\mathrm{dB}) = -P_L = 10\log_{10}\frac{P_{\mathrm{RX}}}{P_{\mathrm{TX}}} \tag{2.10}$$

The free space path loss and path gain is then:

$$\begin{aligned} P_L(\mathrm{dB}) &= 10\log_{10}\frac{P_{\mathrm{TX}}}{P_{\mathrm{RX}}} \\ &= -10\log_{10}\frac{G_{\mathrm{RX}}G_{\mathrm{TX}}\lambda^2}{(4\pi d)^2} \\ &= -10\log_{10}(G_{\mathrm{TX}}G_{\mathrm{RX}}) - 20\log_{10}(\lambda) + 20\log_{10}(4\pi) + 20\log_{10}(d) \end{aligned} \tag{2.11}$$

$$P_G(\mathrm{dB}) = -P_L = 10\log_{10}\frac{G_{\mathrm{RX}}G_{\mathrm{TX}}\lambda^2}{(4\pi d)^2} \tag{2.12}$$

This free space path loss model is only applied in far-field scenario; in other words, it is valid when d is in the far field of the transmitting antenna. This means the distance d must be greater than the far-field distance, also called Fraunhofer distance, d_f:

$$d_f = \frac{2D^2}{\lambda} \tag{2.13}$$

where D denotes the largest physical linear dimension of the antenna. Moreover, to make sure the receiver is in the far-field region, d_f must satisfy

$$d_f \gg D \tag{2.14}$$

and

$$d_f \gg \lambda \tag{2.15}$$

Apparently, eq. (2.7) does not hold for $d = 0$. For this reason, the simplified free space model chooses a reference point at d_0, $d_0 \geq d_f$, where the $P_{RX}(d_0)$ may be measured or predicated by any model, and the received signal power at any $d \geq d_0$ is related to P_{RX} at d_0.

$$P_{RX}(d) = P_{RX}(d_0)\left(\frac{d_0}{d}\right)^2 \qquad d \geq d_0 \geq d_f \tag{2.16}$$

In practical systems, for low-gain antennas in the 1–2 GHz band, the typical reference distance d_0 is chosen 1 m in indoor environment and 100 m or 1 km in outdoor environment.

2.3 Ray tracing mechanisms and models

A radio signal transmitted in a realistic environment might encounter different kinds of dielectric and conducting obstacles that introduce reflection, diffraction or scattering effects, as shown in Fig. 2.2.

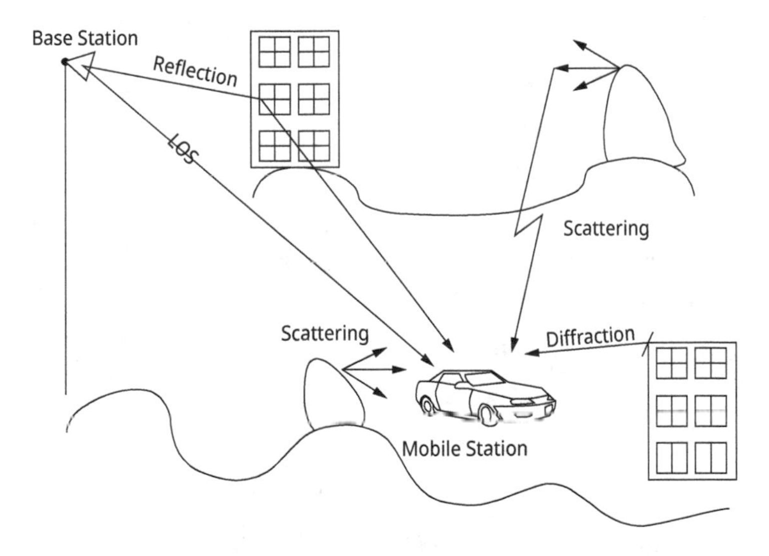

Fig. 2.2: Multipath effects caused by various obstacles.

For an obstacle with a smooth surface, the signal will be mainly reflected, and part of the energy would penetrate into the object. For the obstacle with rough surfaces, the signals are diffusely scattered. The wave can also be diffracted at the shape edges of the obstacles. These components that are reflected, diffracted, or scattered are called multipath components, and they are different from the LOS signal at the receiver since they can be attenuated in power, delayed in time, shifted in phase and/or frequency. The LOS and multipath components are summed up at the receiver. By the ray tracing mechanisms, we assume the number of obstacles in the environment is finite, each with the known location and dielectric properties. Theoretically, the multipath components can be obtained by solving Maxwell's equations with boundary conditions describes the multipath propagation details. However, it is impractical because the computational complexity is too high. By representing the wave fronts as simple particles, the ray tracing models approximate the main propagation mechanisms including reflection, diffraction, and scattering effects using simple geometric equations, thus describing the propagation of electromagnetic waves approximately. The ray trace model approaches the theoretical solution when the distance of receiver and the nearest obstacle beyond many wavelengths and the size of the obstacle are much larger than the wavelength and the surface is fairly smooth. Let's first discuss the three main mechanisms, and then discuss the ray trace models.

2.3.1 Reflection

Reflection may occur multiple times before the electromagnetic wave arrive at the RX. The power of reflected copy of the transmitted signal depends on the reflection coefficient of the surface, and the direction into the surface.

Let's consider a wave that is incident on a dielectric half space. As shown in Fig. 2.3, the angle incidence is equal to the reflected angle:

$$\theta_r = \theta_e \tag{2.17}$$

The reflection coefficient can be computed for transversal electric and transversal magnetic. We only give the special case that the wave is incident from air to ground. In this case, the reflection coefficient is given by

$$R = \frac{\sin\theta - Z}{\sin\theta + Z} \tag{2.18}$$

where

$$Z = \begin{cases} \sqrt{\varepsilon_r - \cos^2\theta}/\varepsilon_r & \text{for vertical polarization} \\ \sqrt{\varepsilon_r - \cos^2\theta} & \text{for horizontal polarization} \end{cases} \tag{2.19}$$

where ε_r is the dielectric constant of the ground, which is approximately a pure dielectric value of about 15 for ground or road, and $\theta = \frac{\pi}{2} - \theta_e$.

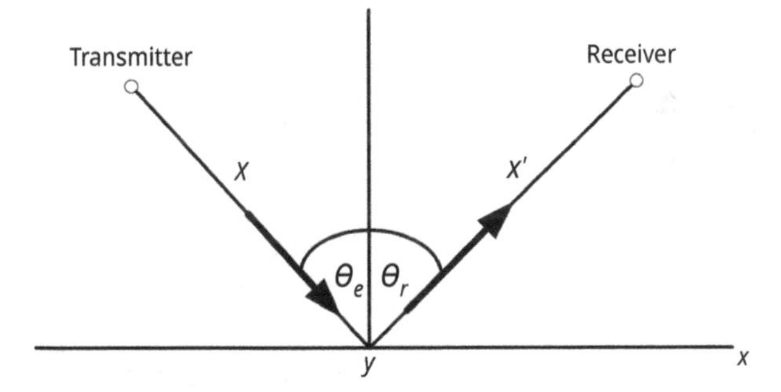

Fig. 2.3: Diagram of reflection.

In the figure, the path length of the reflected signal is $x+x'$, thus producing a phase shift of $\varphi = 2\pi(x+x')/\lambda$.

Thus, we obtain the received reflected signal:

$$r(t) = \mathrm{Re}\left\{R\sqrt{G_{r-\mathrm{TX}}G_{r-\mathrm{RX}}}u(t-\tau)e^{-j2\pi(x+x'/2}e^{-j2\pi f_c t}\right\} \tag{2.20}$$

where $G_{r-\mathrm{TX}}$ and $G_{r-\mathrm{RX}}$ are the transmit and receive antenna gain in the direction of the reflected ray separately, and τ is the delay associated the reflected ray relative the LOS path.

2.3.2 Diffraction

Diffraction occurs when the transmitted signal bend around small or sharp obstacles, or when the transmitted signal spreads out after they pass through small openings in its path to the receiver. Diffraction effect in wireless communications results from the natural phenomena like the curved surface of the earth, hilly or artificial phenomena like building edges, and so on. There are three diffraction models with different complexities. Geometrical theory of diffraction can characterize the diffraction accurately with an impractical complexity. Wedge diffraction model is used to characterize the mechanism by which signals are diffracted for some incident angles on the wedge. It is not commonly used for its requirement for a numerical solution. Fresnel knife edge diffraction model so far is the most commonly used one, as shown in Fig. 2.4.

In this model, we assume the diffracting object be asymptotically thin, and ignore other parameters such as conductivity, surface roughness, and polarization, and this may introduce inaccuracies.

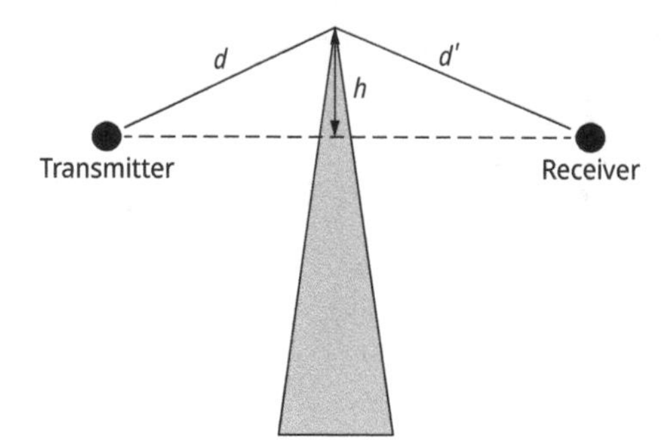

Fig. 2.4: Fresnel knife edge diffraction model.

In the figure, the path length of the diffracted signal is $d+d'$, resulting in a phase shift of $\varphi = 2\pi(d+d')/\lambda$. For smaller h relative to d and d', distance between the path the diffracted signal traveled and the LOS length is approximately to

$$\Delta d \approx \frac{h^2(d+d')}{2dd'} \tag{2.21}$$

and the corresponding phase shift respect to the LOS path is

$$\Delta\phi = \frac{2\pi\Delta d}{\lambda} \approx \frac{\pi}{2}v^2 \tag{2.22}$$

where

$$v = h\sqrt{\frac{2(d+d')}{\lambda dd'}} \tag{2.23}$$

The parameter v is called Fresnel-Kirchhoff diffraction parameter. In this model, the path loss associated with the diffraction is expressed as a function of v. Unfortunately, however, there is typically no closed form expressions for the path loss in diffraction and the complex Fresnel integral are used to find the path loss. An approximations path loss (in dB) relative to LOS path loss can be found in Lee (1982) as follows:

$$L(v)\text{dB} = \begin{cases} 20\log_{10}(.5 - .62v) & -0.8 \le v < 0 \\ 20\log_{10}(.5e^{-0.95v}) & 0 \le v < 1 \\ 20\log_{10}[.4 - \sqrt{.1184 - (.38 - 0.1v)^2}] & 1 \le v < 2.4 \\ 20\log_{10}(.225/v) & v > 2.4 \end{cases} \tag{2.24}$$

Thus, we obtain the received diffracted signal:

$$r(t) = \text{Re}\left\{L(v)\sqrt{G_{d\text{-RX}}G_{d\text{-TX}}}u(t-\tau)e^{-j2\pi(d+d)/2}e^{-j2\pi f_c t}\right\} \tag{2.25}$$

where $G_{d\text{-RX}}$ and $G_{d\text{-TX}}$ are the transmit and receive antenna gain in the direction of diffracted ray, and τ is the delay associated with the diffracted ray relative to the LOS one.

2.3.3 Scattering

Scattering occurs when the surface of the object between the transmitter and receiver is not smooth, as shown in Fig. 2.5. Suppose the scattered ray has two segments s and s', as shown in the figure, and the received scattered ray is expressed by the bistatic radar equation as

$$r(t) = \text{Re}\left\{ u(t-\tau) \frac{\lambda \sqrt{G_{S\text{-TX}} G_{S\text{-RX}}} \sigma e^{-j2\pi(s+s')/2}}{(4\pi)^{3/2} ss'} e^{-j2\pi f_c t} \right\} \tag{2.26}$$

where $\tau = (s + s')/c$ denotes the delay of the scattered ray, σ (in m^2) denotes the radar cross section of the scattering object and the value is determined by the roughness, size, and shape of the scatterer, $G_{s\text{-TX}}$ and $G_{s\text{-RX}}$ are the antenna gains in the considered transmit and receive direction. In this model, we assume that the signal first travels to the scatterer by free space model, and is then reradiated by the scatterer where its transmit power is σ times the received power at the scatterer. The received power is inversely proportional to the product of s and s'.

We can also rewrite the path loss in dB form. The empirical value of 10 log10 σ varies for different buildings, and it ranges from −4.5 to 55.7 dBm^2, where dBm^2 denotes the dB value of the measurement for σ with respect to one square meter.

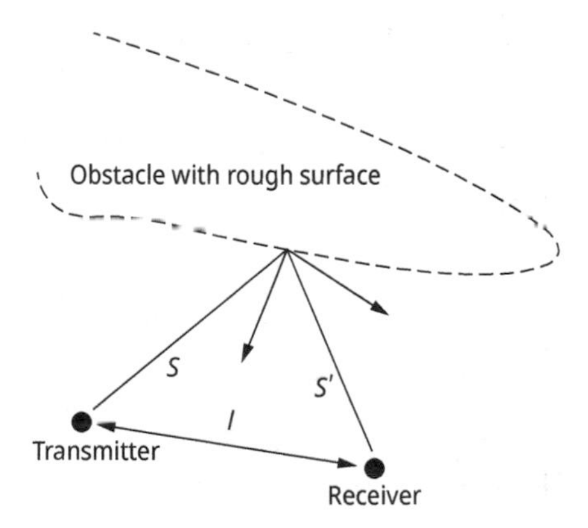

Fig. 2.5: Scattering.

2.3.4 General ray tracing model

If the LOS ray, N_r reflected rays, N_d diffracted rays and N_s scattered rays are all included, the final received signal of a general ray tracing model is represented as follows:

$$\begin{aligned} r_{\text{total}}(t) = \Re\Bigg\{ \frac{\lambda}{4\pi} \Bigg[\sum_{l=0}^{N_l} \frac{\sqrt{G_{\text{TX}}G_{\text{RX}}}u(t)e^{j2\pi l/\lambda}}{l} + \sum_{i=1}^{N_r} \frac{R_{x_i}\sqrt{G_{r\text{-TX}_i}G_{r\text{-RX}_i}}u(t-\tau_i)e^{-j2\pi x_i/\lambda}}{x_i} \\ + \sum_{j=1}^{N_d} L_j(v)\sqrt{G_{d\text{-TX}_j}G_{d\text{-RX}_j}}u(t-\tau_j)e^{-j2\pi\left(d_j+d_j'\right)/\lambda} \\ + \sum_{k=1}^{N_s} \frac{\sqrt{G_{s\text{-TX}_k}G_{s\text{-RX}_k}\sigma_k}u(t-\tau_k)e^{j2\pi\left(s_k+s_k'\right)/\lambda}}{s_k s_k'} \Bigg] e^{j2\pi f_c t} \Bigg\} \end{aligned} \tag{2.27}$$

where N_l may be 0 or 1, and 0 means no LOS ray exists. Any of these path loss components may have an additional attenuation factor if it needs to penetrate obstructions on its path.

To use this model, a detailed geometry and dielectric properties of site are required. The computation is generally performed by using computer packages with a database including the required properties.

2.3.5 Ground reflection two-ray model

Two-ray model is a useful model to predicate large-scale signal strength over distance of several kilometers. It is based on the geometric optics to model two paths from the TX to the RX, where one is the LOS and another is the reflection from the ground, which is illustrated in Fig. 2.6.

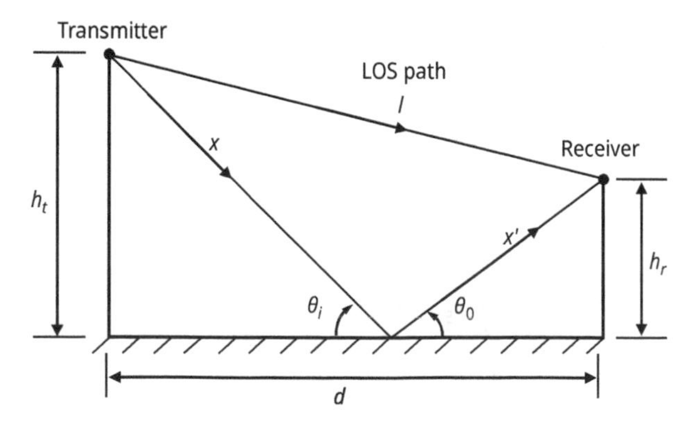

Fig. 2.6: Ground reflection two-ray model.

The received signal can be an LOS ray given by free space model plus a reflected ray given by the following equation:

$$r_{2\text{ray}}(t) = \Re\left\{\frac{\lambda}{4\pi}\left[\frac{\sqrt{G_{\text{TX}}G_{\text{RX}}}u(t)e^{-j2\pi l/2}}{l} + \frac{R\sqrt{G_{r-\text{TX}}G_{r-\text{RX}}}u(t-\tau)e^{-j2\pi(x+x')/\lambda}}{x+x'}\right]e^{j2\pi f_c t}\right\} \tag{2.28}$$

where G_{TX}, G_{RX}, $G_{r-\text{TX}}$ and $G_{r-\text{RX}}$ are the same as those in eqs. (2.7) and (2.20).

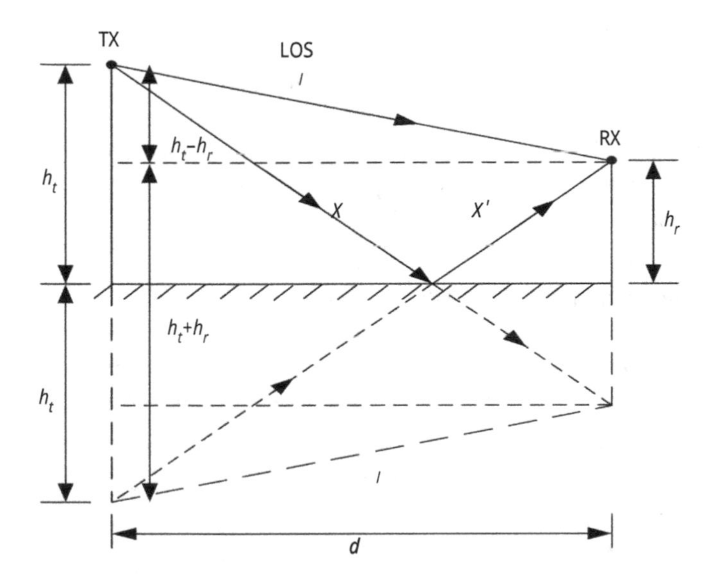

Fig. 2.7: The method of image to determine the length difference of the two rays.

By using the method of image shown in Fig. 2.7, we can see the length distance of the two rays is

$$x + x' - l = \sqrt{(h_t + h_r)^2 + d^2} - \sqrt{(h_t - h_r)^2 + d^2} \tag{2.29}$$

The corresponding phase difference is

$$\Delta\phi = 2\pi(x + x' - l)/\lambda \tag{2.30}$$

When d is very large compared to h_t and h_r, it can be approximated to the following by using a Taylor series:

$$\Delta\phi = \frac{2\pi(x + x' - l)}{\lambda} \approx \frac{4\pi h_i h_r}{\lambda d} \tag{2.31}$$

For the narrowband transmitted signal where the delay spread $\tau \ll 1/B_u, u(t) \approx u(t-\tau)$, and asymptotically large d, $x+x' \simeq l \simeq d$, $\theta \approx 0$, $G_{TX} \approx G_{r-TX}$, $G_{RX} \approx G_{r-RX}$, and $R \approx -1$. The received power of the two-ray model is thus

$$P_r = P_t\left(\frac{\lambda}{4\pi}\right)^2\left|\frac{\sqrt{G_{LX}G_{RX}}}{l} + \frac{R\sqrt{G_{r-TX}G_{r-RX}}e^{-j\Delta\phi}}{x+x'}\right|^2 \approx \left(\frac{\lambda\sqrt{G_{TX}G_{RX}}}{4\pi d}\right)\left(\frac{4\pi h_t h_r}{\lambda d}\right)^2 P_t$$
$$= P_t G_{TX} G_{RX}\frac{h_t{}^2 h_r{}^2}{d^4} \tag{2.32}$$

In dB form, it can be rewritten as

$$P_r = P_t + 10\log_{10}(G_{TX}G_{RX}) + 20\log_{10}(h_t h_r) - 40\log_{10}(d) \tag{2.33}$$

It shows that the received power falls off with the fourth power of distance d when d is large enough. Different from that of the free space path loss model, the path loss is much more rapid and independent of the wavelength λ. For small d, empirical results show that path loss is approximately proportional to d^2. When $d = \frac{4h_t h_r}{\lambda}$, eq. (2.33) is evaluated for $\Delta\phi = \pi$ and the ground appears in the first *Fresnel zone, where the Fresnel zone* denotes the successive regions where distance of the secondary path from the transmitter to receiver is $n\lambda/2$ greater than that of the LOS ray. Apparently, the two paths are added up destructively for the path difference greater than $\lambda/2$ and the path loss is more rapidly. So we call $d_c = \frac{4h_t h_r}{\lambda}$ critical distance. Critical distance is a useful in calculating the microcell path loss and cell size design in the first-generation cellular system.

2.4 Empirical outdoor path loss models

It is difficult for a free space path loss model and ray tracing models to accurately characterize the path loss in irregular environments where most mobile communication systems operate, especially the obstacles such as trees, buildings and other objects that may affect the signal propagation must be considered. Researchers presented a number of empirical path loss models to predict the received power in typical environments. These models are mainly based on empirical measurements over a given distance in a given frequency range for a particular environment. Some of the models may be extended to more general scenarios. We'll discuss the commonly used outdoor path loss models; indoor models will be discussed in the next section.

2.4.1 Okumura path loss model

Okumura model is a widely used empirical path loss model for signal prediction in large urban macrocells (Okumura et al. 1968). The model is suitable for applications with a coverage distance of 1–100 km and a frequency range of 150–1920 MHz. The frequency band can be extrapolated to 3 GHz. The base station heights for this model were 30–1000 m. The model is expressed as

$$P_L(d)\text{dB} = L(f_c, d) + A_{mu}(f_c, d) - G(h_t) - G(h_r) - G_{\text{AREA}} \tag{2.34}$$

where $P_L(d)$ is the path loss as a function of d, $L(fc, d)$ is the path loss at distance d and carrier frequency f_c for free space path loss model, $A_{mu}(fc, d)$ is the median attenuation relative to path loss in free space model across all environments, $G(h_t)$ denotes the antenna height gain factor of the base station, $G(h_r)$ denotes the antenna height gain factor of the mobile station, and G_{AREA} is the gain for the considered type of environment. The antenna height gains are functions of height, and the empirical formulas for $G(h_t)$ and $G(h_r)$ derived by Okumura are

$$G(h_t) = 20\log_{10}(h_t/200), \quad 30\text{ m} < h_t < 1000\text{ m} \tag{2.35}$$

$$G(h_r) = \begin{cases} 10\log_{10}(h_r/3) & h_r \leq 3\text{ m} \\ 20\log_{10}(h_r/3) & 3\text{ m} < h_r < 10\text{ m} \end{cases} \tag{2.36}$$

G_{AREA} for a wide range of frequencies and area can be found in Fig. 3.24 of Rappaport (2002); we give several typical values in Tab. 2.2, and other values can be estimated by interpolation.

Tab. 2.2: Typical G_{AREA} values (dB).

	100 MHz	500 MHz	1 GHz	3 GHz
Open area	22	25	28	35
Quasi-open area	16.7	19	22	29
Suburban area	5	7	10	13

2.4.2 Hata model

The Hata model is another empirical path model applicable for frequency band 150–1500 MHz, and it uses the graphical path loss data provided by Okumura. This is a closed-form formula and is standard formula for empirical path loss in urban areas. It gives correction equations for other situations. The standard Hata formula for urban areas is as follows:

$$P_{L,\text{urban}}(d)\text{dB} = 69.55 + 26.16\log_{10}(f_c) - 13.82\log_{10}(h_t) - a(h_r) + [44.9 - 6.55\log_{10}(h_t)]\log_{10}(d) \tag{2.37}$$

where f_c is the frequency (in MHz) in the region from 150 to 1500 MHz, h_t and h_r are the transmitter antenna height (in meters) and receiver antenna height (in meters) separately: $30\text{ m} \le h_t \le 200\text{ m}$ and $1\text{ m} \le h_r \le 10\text{ m}$. The parameter d is distance between transmitter and receiver (in km), and $a(h_r)$ is the correction factor for the antenna height. For small to medium sized cities,

$$a(h_r) = (1.1\log_{10}(f_c) - .7)h_r - (1.56\log_{10}(f_c) - .8)\text{dB} \tag{2.38}$$

and for a large city, it is given by

$$a(h_r) = 3.2(\log_{10}(11.75h_r))^2 - 4.97\text{ dB} \quad \text{for } f_c \ge 300\text{ MHz} \tag{2.39}$$

$$a(h_r) = 8.29(\log_{10}(1.54h_r))^2 - 1.1\text{ dB} \quad \text{for } f_c < 300\text{ MHz} \tag{2.40}$$

To obtain the path loss in suburban area, the standard Hata formula in eq. (2.37) is modified as

$$P_{L,\text{suburban}}(d) = P_{L,\text{urban}}(d) - 2[\log_{10}(f_c/28)]^2 - 5.4 \tag{2.41}$$

and for path loss in open rural areas, the formula is modified as

$$P_{L,\text{rural}}(d) = P_{L,\text{urban}}(d) - 4.78[\log_{10}(f_c)]^2 + 18.33\log_{10}(f_c) - K \tag{2.42}$$

where k is the constant related to the environment, and its value ranges from 35.94 (rural) to 40.94 (desert).

When $d > 1$ km, the result from the Hata model is very closely to the Okumura model so that it is suitable for early large cellular systems.

2.4.3 Extension to Hata model

The Hata model was extended to 2 GHz by the European Cooperative for Scientific and Technical research (EURO-COST), which is expressed as

$$P_{L,\text{urban}}(d)\text{dB} = 46.3 + 33.9\log_{10}(f_c) - 13.82\log_{10}(h_t) - a(h_r) + (44.9 - 6.55\log_{10}(h_t))\log_{10}(d) + C_M \tag{2.43}$$

where $a(h_r)$ is defined in eqs. (2.38)–(2.40), and C_M takes the value 0 dB for medium-sized cities and suburbs, and takes 3 dB for metropolitan areas.

The COST 231 Extension of the Hata Model is suitable for the following range of parameters:

1.5 GHz < f_c < 2 GHz
30 m < h_t < 200 m
1 m < h_r < 10 m
1 km < d < 20 km

2.5 Indoor path loss models

The radio propagation inside the buildings differs from the traditional radio channel (Cox et al. 1983). Firstly, the distance is generally smaller. Secondly, the environment variation is greater. Although the indoor radio propagation is also dominated by reflection, diffraction and scattering, the materials used in the indoor environments are different greatly in different parts, such as the walls, floors, windows, open areas, the layout of rooms and hallways. These factors affect the path loss in the indoor environment significantly. We will consider these factors in the following sections.

2.5.1 Partition losses in the same floor

Partitions and obstacles inside buildings vary greatly and introduce different losses (Durgin et al. 1998). Average signal loss measurements reported by various researchers for radio paths obstructed by common building material are summarized in the book of Rappaport (2002); typical values for most commonly used environment are cited in Tab. 2.3. Actually, the loss may be a little bit different for different frequencies. In applications that are not particularly precise, we can also ignore this difference.

Tab. 2.3: Typical partition losses for most commonly used environment.

Material type	Loss (dB)
All metal	26
Aluminum siding	20.4
Foil insulation	3.9
Concrete block wall	13
Loss from one floor	20–30
Loss from one floor and one wall	40–50
Metal catwalk/stairs	5
Light textile	3–5
Heavy textile inventory	8–11
Concrete block wall	13–20
Concrete wall	8–15
Concrete floor	10
Commercial absorber	38–59
Aluminum(1/8 in) – 1 sheet	46–53

2.5.2 Partition losses between floors

The path losses between floors are determined by several factors: external dimensions and material of building, materials and structure of floor, number of windows and window types and so on. Measurements show that there is 10–20 dB attenuation loss for 900 MHz signal penetrating a single floor, and then the subsequent floor attenuation loss will be 6–10 dB per floor for the next three floors, but there is a few dB per floor for more than four floors. Typically, the attenuation loss per floor for higher frequency is larger.

2.5.3 Indoor attenuation loss model

The indoor path loss model can be written as

$$P_r\,(\mathrm{dBm}) = P_t\,(\mathrm{dBm}) - P_L(d) - \sum_{i=1}^{N_f} \mathrm{FAF}_i - \sum_{i=1}^{N_p} \mathrm{PAF}_i \tag{2.44}$$

where $P_L(d)$ denotes the path loss from the analytical or empirical model, FAF_i represents the floor attenuation factor (FAF) for the ith floor penetrated by the signal, and PAF_i represents the partition attenuation factor (PAF) associated with the ith partition penetrated by the signal. The number of floors and partitions penetrated by the signal are N_f and N_p, respectively.

For a receiver receiving signal from transmitter outside the building, the building materials must be considered to calculate the penetration loss. For signal at 900 MHz to 2 GHz, the penetration loss on the first floor is typically from 8 to 20 dB and decreases by about 1.4 dB per floor at floors above the first floor (Parsons 1992, De Toledo et al. 1998, Hoppe et al. 1999, Goldsmith 2005). And, of course, if there are windows, the loss will be decreased.

2.6 Log-distance path loss model

Although many analytical and empirical path loss models were proposed, there is no path loss model that can be widely adapted to all environments. It has been shown by many researchers that the average received signal power decreases logarithmically with distance commonly for many environments:

$$P_r(d) = P_t P_L(d_0) \left[\frac{d_0}{d}\right]^n \tag{2.45}$$

In dB form,

$$P_r(d)\,(\text{dBm}) = P_t(\text{ dBm }) + P_L(d_0)(\text{dB}) - 10n\log_{10}\left[\frac{d}{d_0}\right] \tag{2.46}$$

Typical values for n are shown in Tab. 2.4 (Goldsmith 2005, Rappaport 2002, Andersen et al. 1995). Actually, the exponents also relate to frequency bands; interested readers can refer to Andersen et al.'s (1995) work. The parameter d_0 is a reference distance for the antenna far field, 1 km is commonly used in large cells, and 100 m or 1 m may be used for microcell. $P_L(d_0)$ is the free space path loss for distance d_0.

Tab. 2.4: Path loss exponents for different environments.

Environment	*n* Range
Free space	2
Urban macrocells	3.7–6.5
Urban microcells	2.7–3.5
Retail store	1.8
Grocery story	2.2
Shadowed urban cellular radio	3–5
Office building (line of sight)	1.6–1.8
Office building(same floor)	1.6–3.5
Office building with obstruction	4–6
Office building (multiple floors)	2–6
Store	1.8–2.2
Factory	1.6–3.3
Factory with obstruction	2–3
Home	3

2.7 Log-normal shadowing

Besides the path loss effects, the blockage and obstacles may cause random variations to power of the received signal at a given distance. Since the number, the location, size, and dielectric properties of the obstacles are constantly changing and generally unknown, statistical models must be used. The additional attenuation is commonly described by log-normal shadowing, which match the measurements quite well. By taking the shadowing effect into account, the path loss can be rewritten as

$$P_r(d)\,(\text{dBm}) = P_t\,(\text{dBm}) + P_L(d)(\text{dB}) + \psi_{\text{dB}} \tag{2.47}$$

where P_t is the transmitter power, $P_L(d)$ is the path loss obtained by any analytical or empirical model, ψ_{dB} expresses lognormal shadowing effect which is a Gaussian distributed variable (in dB) with zero-mean and standard deviation $\sigma_{\psi_{\text{dB}}}$. Most empirical studies for outdoor channels support a standard deviation $\sigma_{\psi_{\text{dB}}}$ ranging from 4 to 13 dB.

2.8 Link budget

Link budget plays an important role in the planning of wireless systems. A certain minimum transmission quality such as bit error probability is necessary for a wireless system. This minimum transmission quality is corresponding to a minimum signal-to-noise power ratio (SNR) at the receiver. In a scenario where there is only one BS transmitter and one MS receiver, the performance of the system is solely determined by the ratio of the desired signal power and the noise power. If there is an interference source besides the noise, signal-to-interference power ratio should be considered.

2.8.1 Calculation of the noise power

The noise typically includes the following components:

Thermal noise:
Thermal noise depends on the environmental temperature and can be calculated by

$$N_0 = -k_B T_e \tag{2.48}$$

where k_B is Boltzmann's constant, $k_B = 1.38 \times 10^{-23}$J/K. T_e is the environmental temperature where $T_e \approx 300$ K on the Earth, and $T_e \approx 4$ K in the cold sky. The noise power is

$$P_n = N_0 B \tag{2.49}$$

where B is the received signal bandwidth (in Hz). The more common way is to write eq. (2.49) in dB and power P is expressed in units of dBm where 0 dBm is 10 log(P/1 mW). For $T_e = 300$ K,

$$N_0 = -174 \text{ dBm/Hz} \tag{2.50}$$

This means that the noise power within a 1 Hz bandwidth is −174 dBm. The noise power inside the bandwidth B is the

$$(-174 + 10\log_{10}(B))\text{dBm} \tag{2.51}$$

The unit for 10 $\log_{10}(B)$ is dBHz.

Out-of-band emissions
Out-of-band emissions include car ignitions and other impulse sources. At 900 MHz, it can be 10 dB stronger than the thermal noise. At the frequency bands of Universal Mobile Telecommunications System, the noise enhancement caused by man-made noise is about 5 dB in urban environments and 1 dB in rural environments. Though the out-of-band emissions are not necessarily Gaussian distributed, they are assumed to be Gaussian for convenience in system planning and designs.

Other intentional emission sources in the unlicensed bands
Systems operating in the unlicensed bands such as the Industrial, Scientific, and Medical band are allowed to emit electromagnetic radiation as long as they satisfy certain restrictions. This kind of interference is negligible.

Receiver noise
The noise of the amplifiers and mixers in the receiver increases the total noise power. The noise figure F is used to describe the effect, which is defined as

$$F = \frac{\mathrm{SNR}_{i\text{-RX}}}{\mathrm{SNR}_{o\text{-RX}}} \tag{2.52}$$

where $\mathrm{SNR}_{i\text{-RX}}$ denotes the SNR at the RX input (typically after down-conversion to baseband), and $\mathrm{SNR}_{o\text{-RX}}$ denotes the SNR at the RX output.

Noise added in the early stages has more impact than that added in the later stages since the amplifiers have gain. The total noise figure F_{eq} of a cascade of components is

$$F_{\mathrm{eq}} = F_1 + \frac{F_2 - 1}{G_1} + \frac{F_3 - 1}{G_1 G_2} + \cdots \tag{2.53}$$

where F_i and G_i are noise figures and amplifier gains of the individual stages.

2.8.2 Link budget calculation

In the link budget calculation, all equations in dB form connecting the transmit power to the received SNR are generally listed. The following points need to be emphasized:

Path loss model selection
An appropriate path loss model must be chosen according to the environment, application, prior information, and convenience of calculation. It may be analytical model, empirical model, or log-distance path loss, as we discussed from 2.2 to 2.7.

Fading margin consideration
As we will discuss in Chapter 3, wireless systems, especially the mobile systems, suffer fading of the transmission channel. It means that even if the distance is approximately constant, the power computed from the path loss model is only a mean value, because the received power is actually a random variable. If the mean received power is used for the link budget, a fading margin must be added to let the minimum received power meet the requirements in a high percentage. In satellite communications, the fading due to clouds and precipitation may be much bigger in deferent frequency bands, so more link margin should be reserved by considering the climatic zone, the carrier frequency, the elevation angle, season, and time of the day.

Difference in uplinks and downlinks

Typically, the noise figures of MSs are higher than that of the BSs because of the cost consideration. Secondly, since the MSs are powered by relatively low-capacity batteries, they cannot emit high power. Finally, because of the difference in size, the way of antenna diversity will also be different.

Consider the downlink of a TD-SCDMA BS that operates at 2110 *MHz. The mobile phone sensitivity is* −105 *dBm. The maximum value of the transmitter amplifier output power is* 20 *W, the transmitter antenna gain is* 18 *dB. Assume there are totally* 5 *dB loss in cables, combiners, etc. in the transmitter is* 5 *dB and total loss in the receiver is* 2 *dB. The fading margin is* 15 *dB. Assume path loss model* (2.46) *is applied with an antenna far-field reference distance of* 100 *m and exponent n of* 3. *What range can be covered?*

$$P_r(d)(\text{dBm}) = P_t\,(\text{dBm}) + P_L(d_0)(\text{dB}) - 10n\log_{10}\left[\frac{d}{d_0}\right]$$

Solution:

TX side				
–	TX power	P_{TX}	20 W	43 dBm
	Antenna gain	G_{TX}	–	18 dB
	Losses in TX	–	–	−5 dB
	EIRP (equivalent isotropically radiated power)	–	–	56 dBm
RX side				
–	RX sensitivity	P_{min}	–	−105 dBm
	Fading margin	–	–	15 dB
	Required minimum RX power (mean)	–	–	−90 dBm
	Admissible path loss (the EIRP minus the required min. RX power)	–	–	146 dB

We obtain the coverage range by solving eq. (2.46) *to find d,*

$$d = d_0 \times 10^{\frac{146+10\times\log_{10}(1/(\lambda/(4\pi d_0)^2))}{10\times n}} = 100\times 10^{\frac{146-132.44}{10\times 3}} = 283.14\ \text{m}$$

If the receiver sensitivity is not given, the computations at the RX become more complicated. The required SNR should be calculated as the metric.

Problems

2.1 State the main factors causing propagation path loss.

2.2 What is the difference between large-scale propagation path loss and small-scale propagation path loss models?

2.3 How is EIRP related to transmitter power and transmitter antenna gain?

2.4 Given a wireless system that operates at 2.4 GHz with an antenna gain of 10 dB, and a free space path loss model that is assumed, find the transmit power required to obtain a received power of 1 dBm for assuming a distance $d = 100$ m.

2.5 Explain the free space propagation model and derive an expression for the received signal power. Make suitable assumption as necessary.

2.6 What is the delay spread of a two-path propagation model when the distance between transmitter and receiver is $d = 100$ m, the transmitter antenna height $h_t = 10$ m, and the receiver antenna height $h_r = 3$ m?

2.7 Describe the procedures to find the standard deviation of the log-normal shadowing.

2.8 State the propagation characteristics of radio waves in the UHF range.

2.9 If the transmitter power of a BS is 10 W and frequency of transmission is 900 MHz, determine the received signal power in dBm at a distance of 1000 m in free space. Assume that the BS transmitter gain is 10 dB and the mobile station receiver gain is 0 dB.

2.10 If the received signal power at 10 m away from a transmitter with an isotropic antenna is 2 mW, assume that the receiver sensitivity is −100 dBm and path loss exponent is 3. Determine the coverage range if an additional 10 dB margin is needed to compensate for shadowing loss.

Chapter 3
Wireless propagation channel modeling: small-scale fading and statistical multipath channel

Small-scale fading is introduced by the constructive and destructive addition of different multipath components of the channel. We already know that the deterministic multipath effects can be described by ray-tracing models given in Chapter 2. However, deterministic channel models are rarely applicable in actual environments. In practice, multipath channel can be modeled by a random time-varying impulse response. In this way, the important characteristics of a channel can be described by finding the statistical properties of the model.

When a single pulse is transmitted through a multipath channel, a signal-pulse sequence will be obtained at the receiver. Each pulse in the sequence may correspond to the line-of-sight (LOS) path or a distinct multipath component, and each multipath component is associated with a distinct scatter or cluster of scatters. We used time-delay spread to describe the channel which is defined as the time delay between the arrival of the first received pulse (LOS or multipath) and the last received multipath component.

If the time delay spread of the channel is smaller than the inverse of the signal bandwidth, the time spreading in the received signal is also small. However, the multipath components added up together also cause fast variation in the received signal strength. It is necessary to characterize the small-scale variations due to the constructive and destructive addition of multipath components. For the relatively large delay spread, the time spreading of the received signal is significant and substantial signal distortion may be introduced. Characterizing the statistical properties of wideband multipath fading channels needs two-dimensional transforms based on the underlying time-varying impulse response.

3.1 Factors influencing small-scale fading

In the radio propagation environment, there are several typical physical factors that influence small-scale fading, which can be figure out in the following:

3.1.1 Multipath propagation and the signal bandwidth

The signal transmission environment produced by signal propagation effects such as reflection, scattering and diffraction makes signal energy dissipated in amplitude, phase and time. The multipath mechanisms are different for different system. For example, for HF radio, the multipaths are mainly reflection from multiple ionosphere layers, whereas for radio LAN, the multipaths are mainly due to reflection from walls

https://doi.org/10.1515/9783110751437-003

and building structure. Table 3.1 lists some of the mechanisms. These effects result in random number of copies of transmitted signal at the receiving antenna with displacements. The randomness of the phase and amplitudes of different multipath components yield fluctuations in the signal strength, and this produces small-scale fading and/or signal distortion. If the signal spread of the baseband signal exceeds the inverse of the signal bandwidth, or in other words, the signal bandwidth is greater than the "bandwidth" of the multipath channel, the signal distortion introduces intersymbol interference (ISI). The bandwidth of the multipath channel is defined by the channel coherence bandwidth which is the maximum frequency separation for which signals are strongly correlated in amplitude.

Tab. 3.1: Multipath mechanism in wireless communication systems.

	System	Multipath mechanisms
1	Mobile and personal radio	Reflection and scattering from buildings, terrain, etc.
2	Microwave point-to-pint link	Atmospheric refraction and reflection
3	Satellite-mobile systems	Ground and building reflection
4	Diffuse infrared	Reflection from walls

3.1.2 Relative motion of the receiver, transmitter and/or surrounding objects

Since different multipath components might have different Doppler shifts, the relative motion between the receiver and the transmitter may result in random frequency offset. The Doppler shift would be positive when the relative motion direction is toward each other, and it would be negative when the relative motion direction is away from each other. If obstacles existed radio channel are in motion, they yield a time-varying Doppler shift on multipath components. If the motion rate of the obstacles is far greater than that of the transmitter or receiver, it will dominate the small-scale fading effect. When the time separation of two signals is within channel coherence time which is impacted by the Doppler shift and will be discussed later, we say the channel is in "static" state.

3.2 Channel impulse response model

Simulating or analyzing radio transmission through the channel is very important in designing wireless systems. Since the wireless channel can be presented as a liner filter with a time-varying impulse response, and the time variation is due to the motion of receiver, transmitter or surrounding objects, the information for simulation or analysis of the channel can be found or derived for the channel impulse response. The filtering nature of the channel is due to the summation of amplitudes and delays of the multiple arriving paths at any time.

Let's use the same expression for the transmitted signal as that in Chapter 2:

$$\begin{aligned} s(t) &= \Re\{u(t)e^{j2\pi f_c t}\} \\ &= \Re\{u(t)\}\cos(2\pi f_c t) - \Im\{u(t)\}\sin(2\pi f_c t) \end{aligned} \tag{3.1}$$

where $u(t)$ is the complex envelope of $s(t)$ with bandwidth B_u and the carrier frequency of $s(t)$ is f_c. The received signal is the sum of signal components including the LOS path and all multipath resolvable:

$$r(t) = \Re\left\{\sum_{i=0}^{N(t)} \alpha_n(t)u(t-\tau_n(t))e^{-j2\pi f_c(t-\tau_n(t)+\varphi_{Dn})}\right\} \tag{3.2}$$

In order to simplify the analysis, we omitted a noise term in the equation above. The path number $i = 0$ corresponds to the LOS path and there is no LOS path if i starts from 1. The number of resolvable multipath components $N(t)$ as well as the path length $r_n(t)$, delay $\tau_n(t) = r_n(t)/c$, Doppler phase shift $\varphi_{D_n}(t)$ and amplitude $\alpha_n(t)$ of each path is generally random. The time-varying channel impulse response is produced from these random components changing with time. For the multipath component produced from one single reflector, the corresponding amplitude $\alpha_n(t)$ is determined by the path loss and shadowing of that multipath component, and the phase variation is associated with delay $\tau_n(t)$ is $\exp[-j2\pi f c\tau_n(t)]$. The corresponding Doppler phase shift is $\phi_{D_n} = \int_t 2\pi f_{D_n}(t)dt$, where $f_{D_n} = v\cos\theta_n(t)/\lambda$ and $\theta_n(t)$ is its angle of arrival.

In eq. (3.2), the nth multipath may from one reflector or a reflector cluster. Two multipath components with delay τ_1 and τ_2 are resolvable if their delay difference significantly exceeds the inverse signal bandwidth: $|\tau_1 - \tau_2| \gg 1/B_u$. Otherwise, the two components cannot be separated at the receiver because $u(t-\tau_1) \approx u(t-\tau_2)$, and they are called nonresolvable components. The nonresolvability of multipaths does not imply that the reflectors must be together or close. By the laws of geometry, the sum of the distances from a point on an ellipse to its two foci is always exactly equal. Assuming that the transmitter and receiver are at the two foci of the ellipse, then the multipath caused by reflectors on the ellipse will always have the same runtime. Of course, in a real environment, these reflectors are not likely to be precisely located on an ellipse (or ellipsoid). According to the bandwidth of the signal, when the runtime difference is far less then inverse of the bandwidth, these reflectors can roughly be regarded as being on an ellipsoid. When a multipath component is a combination of nonresolvable components with approximate the same delay $\tau \approx \tau_1 \approx \tau_2$ the amplitude and phase are the sum of the different components and therefore undergo fast variation due to constructive and destructive combining of these multipath. There are typically two cases. In a narrowband channel, all multipath components are nonresolvable. In a wideband channel, each term in eq. (3.2) corresponds to resolvable multipath component due to a single reflector or multipath components combined together.

Let the time-varying channel impulse response is presented $c(\tau, t_i)$, the received signal $r(t)$ is then represented as a convolution of $s(t)$ with $c(\tau, t)$:

$$r(t) = \int_{-\infty}^{\infty} c(t,\tau)s(t-\tau)d\tau = \Re\left\{\left(\int_{-\infty}^{\infty} c(t,\tau)u(t-\tau)d\tau\right)e^{j2\pi f_c t}\right\} \tag{3.3}$$

For a causal system, $c(\tau, t) = 0$ for $t < 0$, thus eq. (3.3) is reduced to

$$r(t) = \int_{-\infty}^{t} c(t,\tau)s(t-\tau)d\tau = \Re\left\{\left(\int_{-\infty}^{t} c(t,\tau)u(t-\tau)d\tau\right)e^{j2\pi f_c t}\right\} \tag{3.4}$$

There are two time-related parameters in $c(\tau, t)$, t and τ. The parameter t denotes the observation time at the receiver that the impulse response is observed, and the time $t-\tau$ denotes the time that the impulse is put into the channel relative to t. $c(\tau, t) = 0$ means that there is no multipath component with a delay $\tau_n(t) = \tau$ caused by a physical reflector. Suppose $c(\tau, t)$ is a time-invariant channel response and $c(\tau, t) = c(\tau, t+T)$, it means that the response at time t to an impulse at time $t-\tau$ equals the response at time $t+T$ to an impulse at time $t+T-\tau$. Let's set $T = -t$, then $c(\tau, t) = c(\tau, t-t) = c(\tau)$, and we say $c(\tau)$ is the standard time-invariant channel impulse response which means that the response at time τ to an impulse at zero or, equivalently, the response at time zero to an impulse at time $-\tau$.

The three main parameters $\alpha_n(t)$, $\tau_n(t)$ and $\varphi_{D_n}(t)$ related to each of the resolvable multipath components are generally changing with time, so they are modeled as random processes by assuming them as stationary and ergodic. The assumption also holds for the received signals. When the channel is wideband and each term in eq. (3.2) corresponds to a single reflector, these parameters change slowly because they are associated to the propagation environment. When the channel is narrowband, each term in eq. (3.2) corresponds to a summation of nonresolvable multipath components and thus might make these parameters change rapidly on the order of a signal wavelength. Due to the quick changes of each of these parameters made by addition of the different components, they may be constructive and/or destructive.

Comparing eq. (3.4) with eq. (3.2), we obtain

$$c(\tau,t) = \sum_{n=1}^{N} \alpha_n(t)e^{-j\left(2\pi f_c \tau_n(t) + \varphi_{D_n}\right)}\delta(\tau - \tau_n(t)) \tag{3.5}$$

We can simplify $c(\tau, t)$ by letting

$$\phi_n(t) = 2\pi f_c \tau_n(n) - \varphi_{D_n} \tag{3.6}$$

Then eq. (3.6) can be written as

$$c(\tau,t) = \sum_{n=1}^{N} \alpha_n(t)e^{-j\phi_n(t)}\delta(\tau - \tau_n(t)) \tag{3.7}$$

For time-invariant case, the time-varying parameters in $c(\tau, t)$ become constant, and $c(\tau, t) = c(\tau)$ is just a function of τ:

$$c(\tau) = \sum_{n=0}^{N} \alpha_n e^{-j\phi_n \delta(\tau-\tau_n)} \tag{3.8}$$

for discrete multipath components and

$$c(\tau) = \alpha(\tau) e^{-j\phi(\tau)} \tag{3.9}$$

for channels with a continuum of multipath components. In stationary channel case, the response to an impulse at time t_1 is a shifted version of the response to an impulse at time t_2, $t_1 \neq t_2$.

It is easy to understand that $f_c\tau_n(t) \gg 1$ for carrier frequencies of typical wireless systems. For example, for a system with carrier frequency $f_c = 1$ GHz and the delay $\tau_n = 50$ for the nth multipath component (a typical value for an indoor system), $f_c\tau_n = 50 \gg 1$. Since the multipath delays are much greater than 50 ns for outdoor wireless systems, $f_c\tau_n = 50 \gg 1$ also hold. This rapid change of multipath component phase leads to the constructive and destructive addition of multipath components in the received signal, which will make the received signal strength change rapidly. This phenomenon is called *fading*. This fading occurs in the small scale (in the order of signal wavelength); that is why it is also called *small-scale fading*.

The multipath effect on the received signal can be classified into two categories, which is determined by the relative size of the time delay spread of channel and the inverse of signal bandwidth. For typically small channel delay spread scenario, all the rays including the LOS and the multipath components are nonresolvable, and the channel is called narrowband fading channel which will be treated in the next section. For large channel delay spread scenario, all rays including the LOS and the multipath components are typically resolvable, and the channel is called wideband fading model. In wireless systems that the signal are modulated signal, the channel delay spread is typically measured relative to the component according to which one the demodulator is synchronized. Of course, in most cases, the LOS component has the smallest delay τ_0 and the biggest strength, the demodulator synchronizes it, and this results a delay spread which is $T_m = \max_n|\tau_n - \tau_0|$. However, in the case that the demodulator synchronizes to a multipath component with delay equal to the mean delay τ, the delay spread is then represented by $T_m = \max_n|\tau_n - \tau_0|$. In time-varying channels where the multipath delays are changing as a function of time t, the delay spread T_m is a random variable. Moreover, it is really an impossible task to determine the total number of multipath because some multipath components have significantly lower power than others; for example, the multipath that are reflected, scattered or diffracted multiple times, and it is not clear whether these components should be used in the characterization of delay spread. Generally, those components whose power is below the noise floor contribute insignificantly to the delay spread. It is not important to obtain the exact value of delay spread, and this parameter is used roughly to measure the delay associated with significant multipath components. The delay spread of a channel depends highly on the propagation environment. In indoor environment, the typical value is from 10 to 1000 ns; in suburbs it ranges from 200 to 2000 ns and in urban areas the value becomes 1–30 μs.

3.3 Small-scale fading mechanism and channel classification

As shown in Fig. 3.1, the small-scale fading channel can be described from different point of view (Sklar 1997a). From the time delay point of view or from the frequency domain, the channel can be classified into frequency-selective channel and flat fading channel. From the Doppler shift point of view or from time domain, the channel can be classified into two categories: fast fading and slow fading. In actual mobile communications, the signals typically experience small-scale fading superimposed on large-scale fading.

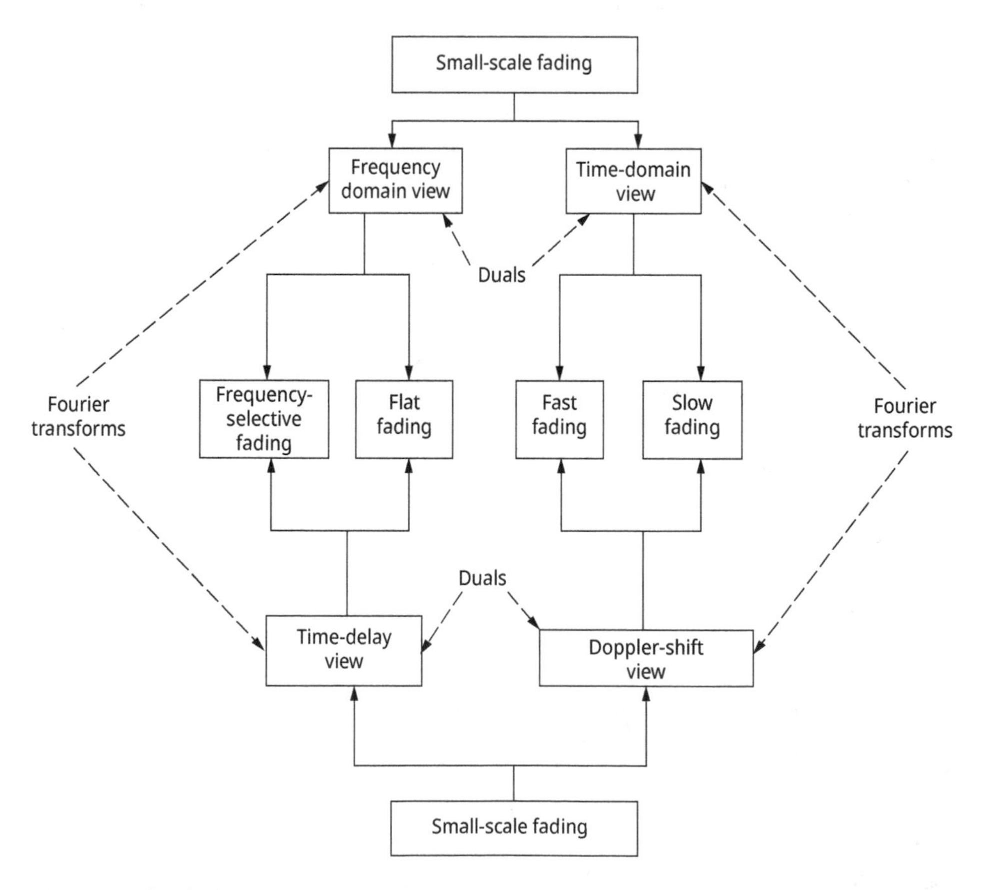

Fig. 3.1: Small-scale fading channel manifestations.

3.3.1 Time delay spreading of a channel

As stated in Section 3.2, $T_m = \max_n|\tau_n - \tau_0|$ is a random variable, the *mean delay spread, rms delay spread* or *delay spread deviation* can be used to describe the delay spread. These parameters cannot be measured directly but can be determined from power delay profile.

Power delay profile

Suppose a short transmitted pulse of duration T is put into a wireless channel and the time delay spread is T_m, then the result received signal would be a pulse of duration $T + T_m$. For linear modulations, information corresponding to a data bit or symbol is carried in a train of pulses in the amplitude and/or phase. In case of $T_m \ll T$, the multipath components are received roughly on top of one another. The resulting constructive and destructive addition of the multipath causes narrowband fading of the pulse, but the time-spreading is very small and will not interfere with subsequently transmitted pulse. In the case when the multipath delay spread $T_m \gg T$, there are different resolvable multipath components that interfere with subsequently transmitted pulses. This effect is called ISI.

Suppose the channel impulse response $c(\tau, t)$ given by eq. (3.7) is a random process. According to the Central Limit Theorem, we assume that $c(\tau, t)$ is a complex Gaussian process for large number of multipath components cases. The Gaussian process is typically characterized by the mean, autocorrelation and cross-correlation.

The autocorrelation function is defined as

$$A_c(\tau_1, \tau_2;t, \Delta t) = E[c^*(\tau_1;t)c(\tau_2;t + \Delta t)] \tag{3.10}$$

For wide-sense stationary (WSS) channels, the above equation reduces to

$$A_c(\tau_1, \tau_2;\Delta t) = E[c^*(\tau_1;t)c(\tau_2;t + \Delta t)] \tag{3.11}$$

In practice, when two components with different delay $\tau_2 \neq \tau_1$ are produced by different obstacles, they are tended to be uncorrelated, and this channel is called WSS uncorrelated scattering; for this kind channel, the autocorrelation becomes

$$E[c^*(\tau_1;t)c(\tau_2;t + \Delta t)] = A_c(\tau_1;\Delta t)\delta[\tau_1, \tau_2] \triangleq A_c(\tau;\Delta t) \tag{3.12}$$

Then, the power delay profile, $A_c(\tau)$, can be defined as the autocorrelation of channel impulse eq. (3.12) with $\Delta t = 0$:

$$A_c(\tau) \triangleq A_c(\tau, 0) \tag{3.13}$$

The power delay profile gives the average power in a certain multipath delay. Comparing to measure the multipath delay directly, it is easier. The average and rms delay spread can be defined by $A_c(\tau)$ as

$$\mu_{Tm} = \frac{\int_0^\infty \tau A_c(\tau)d\tau}{\int_0^\infty A_c(\tau)d\tau} \tag{3.14}$$

$$\sigma_{Tm} = \sqrt{\frac{\int_0^\infty (\tau - \mu T_m)^2 A_c(\tau)d\tau}{\int_0^\infty A_c(\tau)d\tau}} \tag{3.15}$$

The probability density function (pdf) p_{Tm} of the random delay spread T_m can also be defined in terms of $A_c(\tau)$ as

$$p_{Tm}(\tau) = \frac{A_c(\tau)}{\int_0^\infty A_c(\tau)d\tau} \tag{3.16}$$

Apparently, by using this pdf, μ_{T_m} and σ_{Tm} are the mean and rms values of T_m, respectively, with power delay profile which represents the power as the weights so that the stronger the multipath component, the more contribution it makes to the delay spread. Those components whose power below the noise floor will not contribute to the delay spread significantly.

The delay spread of the channel can be roughly described by the delay T when the power delay profile $A_c(\tau) \approx 0$ for $\tau \geq T$. This value T can be used to stand for the delay spread. And it is often taken to be several time of the rms delay spread, for example, $A_c(\tau) \approx 0$ for $\tau > 3\sigma_{Tm}$.

In linear modulation systems, when the rms delay spread $\sigma_{T_m} \gg T_s$, where T_s is the symbol duration time of the modulated signal, the modulated signal will experience significant ISI. When $\sigma_{Tm} \ll T_s$, the ISI is negligible. When $\sigma_{Tm} \approx T_s$, there are some ISI but may not affect the performance significantly.

According to the block time delay view in Fig. 3.1, when the channel rms delay spread $\sigma_{Tm} \gg T_s$, the channel is regarded as frequency-selective fading channel or wideband channel; when $\sigma_{Tm} \ll T_s$, the channel is flat fading channel or narrow band channel.

3.3.2 Channel coherence bandwidth

According to the Fourier transform relationship between time delay view and frequency domain view, another parameter to characterize the channel, the channel coherence bandwidth B_c can be derived from the rms delay spread. Channel coherence bandwidth is a range of frequency that two signals within the bandwidth pass the channel will experience relatively similar fading. And two signals with frequency separation greater than B_c pass the channel will experience quite different fading. The channel coherence bandwidth is approximately equal to

$$B_c = \frac{1}{k\sigma_{Tm}} \tag{3.17}$$

where k is constant related to the frequency correlation coefficient. For example, $B_c \approx 1/5\sigma_{Tm}$ approximates the frequency band over which this correlation exceeds 0.5.

Since delay spread can be found from power delay profile, the channel coherence bandwidth can be derived from the Fourier transform of power delay profile.

Taking Fourier transform of $c\,(\tau, t)$ with respect to τ, we get

$$C(f;t) = \int_{-\infty}^{+\infty} c(\tau;t)e^{-j2\pi f\tau}d\tau \tag{3.18}$$

According to the Fourier transform property, $C(f;\,t)$ would be same as $c(\tau;\,t)$ to be a zero-mean Gaussian random process; therefore it can be characterized by its autocorrelation. $C(f;\,t)$ would be WSS which is same as $c\,(\tau;\,t)$, so that

$$C(f;t) = \int_{-\infty}^{+\infty} c(\tau;t)e^{-j2\pi f\tau}d\tau \tag{3.19}$$

It can be simplified as

$$\begin{aligned} A_C(f_1, f_2;\Delta t) &= E\left[\int_{-\infty}^{+\infty} c^*(\tau_1;t)e^{j2\pi f_1\tau_1}d\tau_1 \int_{-\infty}^{+\infty} c(\tau_2;t+\Delta t)e^{-j2\pi f_2\tau_2}d\tau_2\right] \\ &= \int_{-\infty}^{+\infty}\int_{-\infty}^{+\infty} E[c^*(\tau_1;t)c(\tau_2;t+\Delta t)]e^{j2\pi f_1\tau_1}e^{-j2\pi f_2\tau_2}d\tau_1 d\tau_2 \\ &= \int_{-\infty}^{+\infty} A_c(\tau,\Delta t)e^{-j2\pi(f_2-f_1)\tau}d\tau \\ &= A_C(\Delta f;\Delta t) \end{aligned} \tag{3.20}$$

We define the *spaced frequency correlation function* $A_C(\Delta f) \triangleq A_C(\Delta f;0)$ then

$$A_C(\Delta f) = \int_{-\infty}^{+\infty} A_C(\tau)e^{-j2\pi\Delta f\tau}d\tau \tag{3.21}$$

So the *spaced-frequency correlation function* $A_C(\Delta f)$ and the *power delay profile* are Fourier transform pairs. Since $A_C(\Delta f)$ is defined as an autocorrelation, $A_C(\Delta f) \approx 0$ indicates that the channel response is approximately independent at frequency separation Δf. The channel *coherence bandwidth* corresponds to B_c where $A_C(\Delta f) \approx 0$ for all $\Delta f > B_c$.

In practice, the channel coherence bandwidth can be measured by putting two sinusoid signals with a frequency separation Δf into the channel and calculating the autocorrelation at the receiver. By increasing the frequency separation Δf, we find the bandwidth B_c so that the autocorrelation approach to zero when the $\Delta f > B_c$.

For a modulated signal, if the signal bandwidth is less than the channel coherence bandwidth, the signal pass through the channel will experience similar effect either there is fading or no fading. In this case, we can say the signal experiences flat fading. If the signal bandwidth is far greater than the channel coherence bandwidth, different frequency components of the signal experience different fading. This channel is called frequency-selective fading channel.

3.3.3 Flat fading channel

Flat fading occurs when the delay spread is far less than the symbol period for a linear modulated signal, or more generally, the delay spread is far less the inverse of signal bandwidth. Equivalently, when the signal bandwidth is far less than the channel coherence bandwidth, the channel is called flat fading channel.

In flat fading channel, since $\sigma_{Tm} \ll T_s$, all the multipath components of a symbol arrive at the receiver at approximately the same time; thus they are irresolvable. Therefore, no channel-induced ISI distortion occurs since the fairly small signal time spreading does not cause significant overlap among neighboring received symbols. These irresolvable multipath components may add up together destructively so that a substantial reduction in signal-to-noise (SNR) is caused and the performance is significantly degraded. The fades caused by flat fading channel may go as deep as 20–30 dB so that a very big fade margin is needed to maintain a low bit error rate. The design of the radio link depends heavily on the distribution of the instantaneous gain of the fading channel. Rayleigh distribution is the most common distribution in mobile communications. In Rayleigh flat fading channel, we assume the amplitude varies with time according the Rayleigh distribution.

3.3.4 Frequency-selective fading channel

The condition for frequency-selective fading can also be described in two points of view. In linear modulations, when the delay spread is greater than the symbol period $\sigma_{Tm} > T_s$, or more generally, the delay spread is greater the inverse of signal bandwidth, $\sigma_{Tm} > 1/B$, the multipath components of a symbol arrive at the receiver are resolvable and will interference adjacent symbols. This effect is called ISI, and the corresponding channel is called frequency-selective fading channel. Equivalently, frequency-selective fading channel can also be defined as the channel when the signal bandwidth is greater than the channel coherence bandwidth.

For frequency-selective channel, the channel gain varies with the frequency band. The signals of two frequency components separated over a frequency band greater than the channel coherence experiences different fading when they pass through the channel, and fading is regarded independent.

In modern wireless communications, most of the systems operate in frequency-selective channels. For example, $\sigma_{Tm} \approx 25$ ns is the typical value for indoor environment whereas $\sigma_{Tm} \approx 25$ μs is the typical value for outdoor environment. Let's find the maximum symbol rate $R_s = 1/T_s$ for these situations such that a linearly modulated signal transmitted through these channels experiences negligible ISI. As we know negligible ISI requires $T_s \gg \sigma_{Tm}$, this equivalent to that the symbol rate $R_s = 1/T_s \leq 1/\sigma_{Tm}$. For indoor environment $\sigma_{Tm} \approx 25$ ns, these requirements yield $R_s \leq 4$ Msps, and for binary modulations such as BPSK, $R_b \leq 4$ Mbps. For outdoor environment $\sigma_{Tm} \approx 25$ μs, these requirements yield $R_s \leq 4$ ksps, and for binary modulations, $R_b \leq 4$ kbps. Note that 4 Mbps for indoor systems and 4 kbps for outdoor systems are too low data rate. In order to overcome the performance degradation of linear modulated system caused by ISI, some form of ISI mitigation is needed. We will discuss these techniques in the later chapters. Moreover, ISI is less severe in indoor environments than in outdoor environments because the delay spread values are smaller environments.

3.3.5 Doppler spread and channel coherence time

Delay spread describes the time-dispersive nature of the channel, and equivalently channel coherence bandwidth describes the variation in frequency domain. How the channel time variation nature can be described due to relation motion of transmitter and receiver or movement of objects in the environment? We need two other parameters Doppler spread and channel coherence time to characterize the time varying property of the channel in small scale.

Doppler spread B_D describes the spectral broadening due to the time variation of the mobile channel and is defined as the frequency range over which the Doppler spectrum is nonzero. When a single sinusoidal signal f_c is transmitted over the channel, the received signal may have components in the range $f_c - f_d$ to $f_c + f_d$, where f_d is the Doppler shift. If the bandwidth of the signal is less than B_D, the effect of Doppler spread would be significant and the channel is called fast fading channel. If the bandwidth of the signal is far greater than B_D, the effect of Doppler spread would be negligible and the channel is called slow fading channel.

Doppler spread can also be viewed in Doppler power spectrum. By taking the Fourier transform of $A_C(\Delta f;\Delta t)$ relative to Δt:

$$S_C(\Delta f;\, \rho) = \int_{-\infty}^{+\infty} A_C(\Delta f;\Delta t) e^{-j2\pi\rho\Delta t} d\Delta t \tag{3.22}$$

We set Δf to zero and define the *Doppler power spectrum* of the channel $S_C(\rho) \triangleq S_C(0;\rho)$. It is easily seen that

$$S_C(\rho) = \int_{-\infty}^{+\infty} A_C(\Delta t) e^{-j2\pi\rho\Delta t} d\Delta t \tag{3.23}$$

where

$$A_C(\Delta t) \triangleq A_C(\Delta f = 0; \Delta t) \tag{3.24}$$

is defined as *spaced-time correlation function.*

According the definition, the *Doppler power spectrum* is the Fourier transform of an autocorrelation, and it characterizes the PSD of the received signal as a function of Doppler ρ. The *Doppler spread* of the channel B_D is defined as the maximum value of ρ for which $|S_C(\rho)|$ is greater than zero. Since $A_C(\Delta t)$ and $S_C(\rho)$ are Fourier transform pair, $B_D \approx 1/T_C$, where T_C is the time range over which $A_C(\Delta t)$ greater than zero. When two signals transmitted over time interval is greater than T_C, the fading they experienced can be regarded independent. In general $B_D \approx k/T_C$, where k depends on the shape of $S_C(\rho)$.

3.3.6 Slow fading channel

From Section 3.3.5, we know that slow fading can be viewed in two aspects: In time domain, when the symbol time T_s is far less than the channel coherence time T_c, or in the frequency domain, when the baseband signal bandwidth B_s is far greater than the Doppler spread of the channel B_D, the channel is slow fading channel. The relative motion speed between the transmitter and receiver (or the motion speed of the objects in the channel) and the baseband signal bandwidth determines whether the channel is slow fading or not.

In a slow fading channel, a number of successive symbols experience similar fading. If there is a deep fading, it is likely to cause busty errors. In this case, techniques such as interleaving to randomize the errors are needed.

3.3.7 Fast fading channel

Similar to slow fading, fast fading can also be viewed in two aspects: in time domain, when the symbol time T_s is greater than the channel coherence time T_c, or in the frequency domain, when the baseband signal bandwidth B_s is smaller than the Doppler spread of the channel B_D, the channel is fast fading channel. This fading can also be called *time-selective fading*, and the Doppler spreading makes frequency dispersion, thus leading to signal distortion.

It should be mentioned that when we say a channel is slow fading or fast fading, we do not specify it is flat fading or frequency-selective fading. Fast or slow fading only considers the channel change caused by motion. The channel impulse responds for flat fading can be modeled as delta function since there is no time delay. But for flat fading and fast fading channel, the amplitude of the delta function varies faster than the symbol time. For frequency-selective fading and fast fading channel, the amplitude, phase and time delay of each of the multipaths vary faster than the symbol time. Fast fading occurs only in low data rate situations or in high velocity scenario in practice.

3.4 Performance degradations viewed in different domain

In Section 3.3, we discussed the fading mechanisms, channel classification and performance effects. In this section, we discuss the relationship of them. Table 3.2 lists a summary on fading mechanisms, degradation effects and classification criteria (Sklar 1997a, Sklar 1997b), we will explain it in detail in the following subsections.

3.4.1 Small-scale fading mechanisms in quasi-duality view

In Tab. 3.2, we show the small-scale fading in quasi-dual mechanisms. When the behavior caused by one operation with reference to a time-related domain (time or time delay) is identical to the behavior caused by another operation in reference to the corresponding frequency-related domain (frequency or Doppler shift), we call this pair of operations duality (Sklar 1997a, Sklar 1997b). In this table, the behaviors of these operations are not exactly identical to one another in the strict mathematical sense, but similar, thus we denote them as "quasi-duality." Let's see an example in eqs. (3.21) and (3.24). $A_C(\Delta f)$ in eq. (3.21) shows signal dispersion in the frequency domain, and this gives information about the range of frequency over which two frequency components of a received signal have a strongly potential correlation. $A_C(\Delta t)$ in eq. (3.24) shows the fading rapidity in the time domain, and this gives the information about the span of time over which two received signals have a strongly potential correlation. We call them as dual mechanisms. In Tab. 3.2, we labeled the time-spreading mechanism in the frequency domain and the time-variant mechanism in the time domain.

3.4.2 Performance degradations due to time spreading viewed in time delay domain

When the time spreading of the channel T_m is greater than the symbol time T_s, that is, $T_m > T_s$, a channel is said to be frequency-selective fading channel, or we say it is wideband channel. The multipath components of one symbol are dispersed to other symbols so that ISI distortion is caused. This category of fading degradation can also

Tab. 3.2: Small-scale fading mechanisms, degradation effects and classification criteria.

View points \ Mechanisms		Time-spreading due to multipath effects			Time-variant due to motion		
		Fading type	**Performance degradation**	**Classification criteria**	**Fading type**	**Performance degradation**	**Classification criteria**
Quasi-duality pair between time delay domain and Doppler shift domain	Time delay domain	Frequency-selective fading	ISI distortion, pulse mutilation, irreducible BER	Multipath delay spread > symbol time			
		Flat fading	Loss in SNR	Multipath delay spread ≪ symbol time			
	Doppler shift domain				Fast fading	High Doppler, phase lock loop failure, irreducible BER	Channel fading rate > symbol rate
					Slow fading	Low Doppler, loss in SNR, error burst	Channel fading rate ≪ symbol rate
Quasi-duality pair between frequency domain and time domain	Frequency domain	Frequency-selective fading	ISI distortion, pulse mutilation, irreducible BER	Channel coherence BW ≫ symbol rate			
		Flat fading	Loss in SNR	Channel coherence BW ≫ symbol rate			
	Time domain				Fast fading	High Doppler, phase lock loop failure, irreducible BER	Channel coherence time < symbol time
					Slow fading	Low Doppler, loss in SNR, error burst	Channel coherence time > symbol time

be called channel-induced ISI. It is possible for wireless communications over frequency-selective fading channels to take ISI mitigating techniques to improve the performance because many of the multipath components are resolvable by the receiver.

When the time spreading is far less than the symbol time $T_m \ll T_s$, the channel is said to be flat fading channel. In this case, the delay of all the multipath components is similar and all the multipath components of one symbol reach the receiver within the symbol time duration. These components are irresolvable and there is no ISI distortion caused by the channel, due to the small multipath spreading that does not produce significant overlap among neighboring received symbols. Performance degradation also occurs since the irresolvable phase components can add up destructively to cause SNR loss.

3.4.3 Performance degradation due to time spreading viewed in the frequency domain

A channel can also be referred to frequency-selective channel if the channel coherence bandwidth is less than the symbol rate or bandwidth of baseband signal, $B_c < 1/T_s \approx W$, where we assume the symbol rate, $1/T_s$ is approximately equal to the signal bandwidth W. In practical systems, the two values W and $1/T_s$ may be slightly different due to filtering property or modulation forms.

From view point of frequency domain, the frequency-selective fading indicates that a signal's spectral components in different band are not affected equally by the channel. The signal's spectrum components that are not within the coherence bandwidth will be affected differently (independently) compared to those ones that is within the channel coherence bandwidth.

In frequency domain, flat fading occurs in condition $B_c > W$. This indicates that the same fading (e.g., fading or no fading) occurs across all components of the signal's spectral. Flat-fading channels do not introduce ISI distortion; however it may cause loss in received SNR whenever the signal is in fading. To avoid distortion introduced by frequency-selective fading channel, it is required to set the communication system bandwidth as narrowband to insure

$$B_c > W \approx 1/T_s \tag{3.25}$$

If the channel coherence bandwidth B_c is set to be the upper limit of the transmission rate, ISI can be avoided and equalizer is not a necessary component in the receiver.

In flat fading situation, since $B_c \gg W$ (or $T_m \ll T_s$) is satisfied, the fading within the signal bandwidth is similar. However, since the receiver or transmitter is moving and the radio environment is changing, there may still be such times when the received signal exhibits frequency-selective distortion. Since B_c is much larger than W (or T_m much smaller than T_s), this special situation may occur, but the probability is small and location-dependent. Unlike the frequency-selective fading scenario, where frequency-selective fading occurs all the time and is location independent.

3.4.4 Performance degradation due to time variance viewed in the time domain

The channel can be classified into fast fading and slow fading according to the time variation of the channel. When the channel coherence time T_C is less than the symbol time T_S, $T_C < T_S$, the channel is called fast fading. In fast fading channel, the channel only behaves in a correlated manner is a short time comparing to a symbol time. In certain conditions, the fading may cause the channel characteristics change several times within the duration of one symbol, thus making the baseband pulse shape distorted. This distortion of the baseband pulse results in SNR loss and often causes an irreducible error rate. Such distorted pulses may cause failure of phase-locked-loop receivers and make it difficult to define a matched filter.

Slow fading occurs when $T_C \gg T_S$. In this case the channel shows correlation in a long during comparing to the symbol time. The channel may look unchanged during the time in sequence of symbols is transmitted. In a slow-fading channel, the main performance degradation is loss in SNR. Especially when the channel is in slow deep fading, it affects many subsequent bits and causes bursts of errors.

Finally, in a special case of fast fading, $T_C \ll T_S$, the fading is very fast. This fading is averaged out over a bit or symbol time in the demodulator, so the fading exhibits like AWGN and can be neglected.

3.4.5 Performance degradation due to time variance viewed in the Doppler shift domain

In Doppler shift domain, we can understand the spectral broadening by comparing the channel variation and digital keying. For example, a single tone such as $\cos 2\pi f_c t$ $(-\infty < t < \infty)$is impulses (at $\pm f_c$) in the frequency domain with bandwidth of zero. Digital keying operation can be viewed as multiplying the infinite-duration tone by an ideal rectangular function. The frequency domain description of the ideal rectangular function is of the form $(\sin f)/f$. The resulting spectrum of digital keying modulation is obtained by convolving the spectral impulses with the $(\sin f)/f$ function, thus broadening the spectrum. The faster the signaling rate is, the broader the resulting spectrum of the signal will be. The changing of fading channel is like the keying on and off of digital signals to a certain extent. The faster the channel changes its state, the broader spectrum the received signal will have. This explanation is not strict and exact but it helps in understanding the spectral broadening of time varying channels.

In Doppler shift domain, fast fading channel means the symbol rate, $1/T_s$ (approximately equal to the signaling rate or band width W) is less than the fading rate, $1/T_c$ (approximately equal to f_D). That is to say, fast fading is characterized by

$$W < f_D \tag{3.26}$$

Similarly, slow fading channel means the symbol rate $1/T_s$ is far greater than the fading rate $1/T_c$. In order to avoid signal distortion caused by fast fading, we must make sure the channel is qualified to be a slow fading channel by maintaining the symbol rate greater than the channel fading rate:

$$W \gg f_D \tag{3.27}$$

Or equivalently,

$$T_s \ll T_c \tag{3.28}$$

We can see a conflict in designing a wireless communication system. In eq. (3.25), we can set an upper limit on the signaling rate to keep it far less than the channel coherence bandwidth so that to control the signal dispersion, or equivalently, to ensure the channel is in flat fading. Equation (3.27) shows that we need sets a lower limit on the signaling rate so that the fast fading distortion can be avoided. Early HF communication systems transmitted messages at a low data rate such that the channels were generally be regarded as fast fading. Nowadays, most channels can be viewed as slow fading. If the slow fading condition is not satisfied, the Doppler shift will affect the system performance significantly. The Doppler affects the system performance by introducing an irreducible error floor that cannot be overcome by simply increasing SNR. For communication system in slow fading, coherent demodulators that lock onto and track the information signal should suppress the Doppler Effect. In fast fading channel with large Doppler shift f_D, carrier recovery becomes a difficult problem because phase-locked loops with very wideband (relative to the data rate) need to be designed. A frequency-tracking loop instead of phase-locked loop may help to lower the irreducible error floor by using differential minimum-shift keyed modulation, but it cannot completely remove the effect.

The spread factor of the channel

As the multipath time spread of the channel is T_m, and the Doppler spreading of the channel is demoted as B_D, their product $T_m B_D$ is referred as the channel *spread factor*. A channel is said to be under-spread channel if $T_m B_D < 1$. The channel is said to be overspread channel, if $T_m B_D \geq 1$. If the spread factor $T_m B_D \ll 1$, the channel impulse response can be easily measured and used at the demodulator, and this impulse response information can be used in optimizing the transmitted signal. If $T_m B_D > 1$, it would difficult to reliably measure the channel impulse response.

3.5 Envelop distributions of fading channels

3.5.1 Rayleigh fading channel

Let's first assume that the channel is flat fading channel. The delay spread $\max_{m,n}|\tau_n(t) - \tau_m(t)| \ll 1/B$ holds, and $u(t) \approx u(t-t)$. The received signal (3.2) can be written as

$$r(t) = \Re\left\{u(t)e^{j2\pi f_c t}\left[\sum_{n=0}^{N(t)} a_n(t)e^{j\phi_n(t)}\right]\right\} \tag{3.29}$$

The complex scale factor in the square bracket is introduced by the channel. It is easily seen that the factor is independent of the s(t) or, the baseband $u(t)$, since $T_m \ll 1/B$. We choose $u(t) = 1$ to simplify the influence of the signal in describing the complex scale factor. That is

$$s(t) = \Re\{e^{j(2\pi f_c t + \phi_0)}\} = \cos(2\pi f_c t - \phi_0) \tag{3.30}$$

where ϕ_0 is a random phase offset. Apparently, $T_m \ll 1/B$ is satisfied for any T_m.

Then eq. (3.29) is rewritten as

$$r(t) = \Re\left\{\left[\sum_{n=0}^{N(t)} a_n(t)e^{-j\phi_n(t)}\right]e^{j2\pi f_c t}\right\} = r_I \cos 2\pi f_c t + r_Q(t)\sin 2\pi f_c t \tag{3.31}$$

where

$$r_I(t) = \sum_{n=1}^{N(t)} a_n(t)\cos\phi_n(t), \qquad r_Q(t) = \sum_{n=1}^{N(t)} a_n(t)\sin\phi_n(t) \tag{3.32}$$

where $\phi_n(t) = 2\pi f_c \tau_n(t) + \varphi_{D_n}$. For large $N(t)$, $r_I(t)$ and $r_Q(t)$ are jointly Gaussian distributed according to the Central Limit Theorem.

If there is no a dominant LOS, and $a_n(t)$, $\tau_n(t)$ and Doppler frequency $f_{Dn}(t)$ are changing slowly enough and can be simplified over the time intervals of interest: $a_n(t) \approx a_n$, $\tau_n(t) \approx \tau_n$ and $f_{Dn}(t) \approx f_{Dn}$. Then $\phi_n(t)$ can be rewritten as $\phi_n(t) = 2\pi f_c \tau_n - 2\pi f_{D_n} t - \phi_0$. We have discussed in Section 3.2 that $2\pi f_c \tau_n$ can go through a 360° rotation for a small change in τ_n. Φ_n is uniformly distributed in $[-\pi, \pi]$. Thus

$$E[r_I(t)] = E\left[\sum_n a_n \cos\phi_n(t)\right] = \sum_n E[a_n]E[\cos\phi_n(t)] = 0 \tag{3.33}$$

Similarly, $E[rQ(t)] = 0$.

For any two Gaussian random variables with mean value of zero and equal variance of σ^2, if we define $z(t)$ as

$$z(t) = |r(t)| = \sqrt{r_I^2(t) + r_Q^2(t)} \tag{3.34}$$

Then $Z = \sqrt{X^2 + Y^2}$ is Rayleigh-distributed, and the pdf is

$$p(r) = \begin{cases} \frac{r}{\sigma^2} \exp[-r^2/(2\sigma^2)], & 0 \le r \le \infty \\ 0, & r < 0 \end{cases} \tag{3.35}$$

where σ is the predetermined rms value of the received signal and σ^2 is the predetermined time-average power of the received signal.

For Rayleigh fading pdf, we assume that there is no dominate LOS component; thus, Rayleigh fading represents the worst case of fading for the same time-average received signal power.

The received power Z^2 is exponentially distributed that can be written as

$$p_{z^2}(x) = \frac{1}{P_r} e^{-x/P_r} = \frac{1}{2\sigma^2} e^{-x/(2\sigma^2)}, \quad x \ge 0 \tag{3.36}$$

The above result is based on the flat fading assumption. How about the envelop distribution if the channel is a frequency-selective channel? In fact, for a large number of multipath components $N(t)$, the jointly Gaussian distributed $r_I(t)$ and $r_Q(t)$ still hold due to the central limit theorem. If there is no dominate LOS, Rayleigh distribution for envelop and exponential distribution for power will still hold. However, the performance of the communication over frequency-selective fading not only depends on the envelop distribution but also depends on how severe the ISI is.

3.5.2 Ricean fading channel

If there is a dominant signal component, such as a LOS path, $r_I(t)$ and $r_Q(t)$ are not zero-mean. In this case, the distribution of the received signal envelope is Ricean. The final resulted signal is the superimposition of random multipath signal and the dominant signal. If the dominant signal fades away, the Ricean distribution will degenerate to Rayleigh one.

The pdf of Ricean distribution is

$$p(r) = \begin{cases} \frac{r}{\sigma^2} \exp[-(r^2 + A^2)/(2\sigma^2)] I_0\left(\frac{Ar}{\sigma^2}\right), 0 \le r \le \infty \\ 0, r < 0 \end{cases} \tag{3.37}$$

where A^2 is the power in the dominate component, and $2\sigma^2$ is average power of the nondominate multipath components. The function I_0 is the modified Bessel function of first kind and 0th order. The average power in the received Ricean fading channel is $P_r = \sigma^2 + A^2$.

The Ricean distribution is often represented in terms of a fading parameter K, which is

$$K = A^2/(2\sigma^2) \tag{3.38}$$

Or, In dB form,

$$K(dB) = 10\log_{10}(A^2/(2\sigma^2))(dB) \tag{3.39}$$

Substitute A^2 with $KP_r/(K+1)$ and $2\sigma^2$ with $P_r/(K+1)$, the pdf of Ricean distribution is rewritten as

$$p(r) = \begin{cases} \frac{2r(K+1)}{P_r}\exp\left[-\left(K+\frac{(K+1)r^2}{P_r}\right)\right]I_0\left(2r\sqrt{\frac{(K+1)K}{P_r}}\right), 0 \le r \le \infty \\ 0, r < 0 \end{cases} \tag{3.40}$$

From the equation, we can see that K is a measure if fading. With the decreasing of K, the fading becomes more severe. When K becomes zero, the Ricean fading becomes Rayleigh fading.

3.5.3 Nakagami fading channel

Rayleigh fading channel and Ricean fading channel are two popular channel models capturing the physical properties of the wireless channel. However, experiments have shown that not all channels can be captured by the two-channel models. A more general model the Nakagami fading model was proposed. It was initially proposed, for it matched empirical results for short-wave ionospheric propagation. The pdf of Nakagami distribution is

$$p(r) = \frac{2m^m r^{2m-1}}{\Gamma(m)p_r^m}\exp\left[\frac{-mr^2}{P_r}\right], m \ge 5 \tag{3.41}$$

where P_r is the received signal power and Γ is the gamma distribution. The parameter m can be adjusted to fit a variety of situations. Nakagami fading is suitable for describing the signal amplitude at the receiver in the maximum ratio diversity combining (MRC). The MRC of k Rayleigh-fading signals produces a Nakagami signal with $m = k$. MRC combining of m-Nakagami fading signals in k branches results in a Nakagami signal with shape factor mk. Diversity will be discussed in the later chapters. In practice, the matching of Nakagami distribution with some empirical data is better than other models with these data. Nakagami distribution is a general form because it can reduce to simple form for some values of m. For $m = 1$, it reduces to Rayleigh fading, for $m = \frac{(K+1)^2}{(2K+1)}$, it approximates Ricean fading with parameter K and for $m = \infty$, there is no fading, because the received power P_r is constant. The power distribution of Nakagami fading channel is

$$p_{Z^2}(x) = \left(\frac{m}{P_r}\right)^m \frac{x^{m-1}}{\Gamma(m)} \exp\left(\frac{-mx}{P_r}\right) \tag{3.42}$$

3.6 Level crossing rate and average fading duration

The changing rate of the received signal envelop is related to the signal level and relative motion velocity between transmitter and receiver. The level crossing rate and average fade duration are two important statistical parameters for a Rayleigh fading signal, and they are also key factors in the design of channel codes and diversity schemes in mobile communications.

Level crossing rate is defined as the expected rate for the fading envelope, normalized to the local RMS signal level and crossed a specified threshold, which may correspond a certain signal quality or system performance in the downward direction. For Rayleigh fading channels, the average level crossing number in 1 s crossing a specified level L is given by Goldsmith (2005):

$$N_L = \sqrt{2\pi} f_D \rho e^{-\rho^2} \tag{3.43}$$

where f_D is the maximum Doppler frequency given by V/λ, V is the motion velocity of the mobile station and λ is the wavelength for the carrier frequency. ρ is the value of the specified level L, normalized to the local rms amplitude of the fading envelope, that is, L/L_{rms}.

Or

$$N_L = \sqrt{2\pi}(\mathrm{V}/\lambda)\rho \mathrm{e}^{-\rho^2} \tag{3.44}$$

For Ricean fading envelop (Goldsmith 2005),

$$N_L = \sqrt{2\pi(K+1)} f_D \rho e^{-K-(K+1)\rho^2} I_0\left(2\rho\sqrt{K(K+1)}\right) \tag{3.45}$$

Thus, the level crossing rate is a function of the relative motion velocity V between transmitter and receiver. The maximum level crossing rate occurs at 3 dB below the rms level ($\rho = \sqrt{2}/2 = 0.707$), and the crossing rate is generally small at high and low levels.

The ***average fade duration*** is defined as the average period of the time for which the received signal is below a given fade threshold L. Fading duration is defined for which the signal is below a specified threshold value. The reason we use average fade duration is due to the randomness of fade duration. The average fading duration in Rayleigh fading channel can be expressed as (Goldsmith 2005):

$$\bar{T}_Z = \frac{1}{\sqrt{2\pi}}(e^{-\rho^2} - 1)/(f_D\rho) \tag{3.46}$$

where ρ is the value of the specified level L, normalized to the local rms amplitude of the fading envelope, that is L/L_{rms} and f_D is the maximum Doppler frequency given by V/λ.

In the link budget of the mobile communications, a fade margin should be necessarily considered. The average fade duration of a received signal indicates that the bit or symbols represented by the signal might have high probability to be lost during that fade period. The fade duration is primarily determined by the relative motion speed between transmitter and receiver and inversely proportional to the maximum Doppler frequency. The receiver performance can be evaluated by two important factors: the rate the received signal below a given target level L, and the average duration it remains below this level. Another factor is the depth of fading. It is defined as the mean square of the depth divided by the minimum value of the fading signal. Generally, this value is a random variable. In practice, we use the average depth of fading instead. This does help in relating the SNR and the instantaneous BER during fade duration.

Problems

3.1 What is meant by coherence bandwidth? Determine the coherence bandwidth corresponding to a delay spread of 1 μs.

3.2 List different kinds of fading in the delay spreading view and explain how the performance will be affected by the fading. In order to avoid frequency-selective fading, how can we limit the symbol rate of the communication system? List measures you know that overcome the effect of frequency-selective fading.

3.3 List different kinds of fading in the Doppler shift view and explain how the performance will be affected by the fading. In order to avoid fast fading, how can we limit the symbol rate of the communication system? List measures you know that overcome the effect of fast fading.

3.4 Explain why a small change in the delay $\tau_n(t)$ can make the phase term $\phi_n(t) = 2\pi f_c \tau_n(t) + \varphi_{D_n}$ go through a 360° rotation.

3.5 Explain how fading occurs, and what does it mean by small scale fading?

3.6 What are the distinct applications of Rayleigh and Ricean fading channel model characteristic in wireless communications?

3.7 In what kind of situation Nakagami fading occurs? When it can reduce to Rayleigh fading?

3.8 Given a Rayleigh fading channel has a local rms amplitude of the fading envelope 30 dB, if target fade values $P_0 = 15$ dB, compute the average fade duration for the Doppler frequency of 20 Hz,

3.9 What is meant by coherence time? Mention its significance.

Chapter 4
Digital modulations over wireless channels

The ultimate performance limit of a wireless system is determined by the wireless channel and network design. However, it is a very important task to find suitable modulation schemes, codes and signal-processing techniques to guarantee that the performance limits can be approached. In this chapter, we will treat the modulation and demodulation techniques, and focus on the digital modulations.

Modulation is the process that changes the information message into a pattern suitable for transmission. A block diagram of a communication system can be seen in Fig. 4.1 (Molisch 2011). Of course, some components are optional. From this figure, we can roughly understand the role of the modulation and demodulation modules. Other processing modules are treaded in other chapters. For example, multiple accessing is discussed in Chapter 8. Equalization, channel coding and diversity are treated in Chapter 5. For source coding and decoding, interested readers can refer to other books in this area (Gray 1990, Berger et al. 1975).

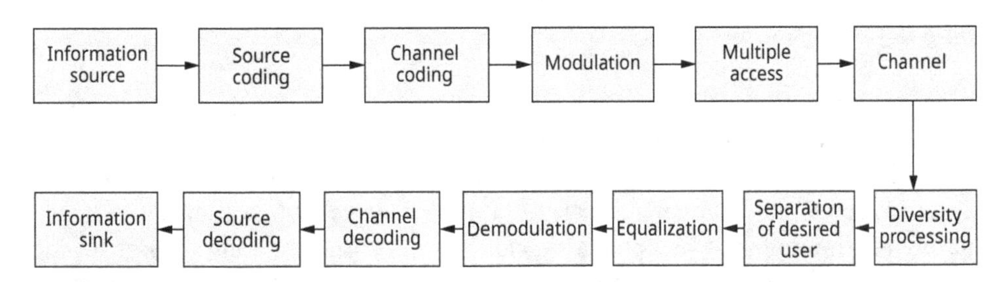

Fig. 4.1: Block diagram of a communication system.

The function of modulation is to modulate the baseband signal to a carrier frequency which is very high compared to the bandwidth of the baseband signal. The information may be conveyed in the amplitude, phase or frequency of the carrier frequency signal, and this will lead to different modulation forms. The inverse process of modulation is called demodulation, which extracts the information from the carrier frequency signal. The extracted information may be further decoded and processed by the receiver.

Modulation can be classified into two main categories: analogy modulation and digital modulation. Nowadays, due to the progress of hardware and digital signal processing, digital transceivers are cheaper, faster and more power-efficient than analog transceivers. Besides, digital modulation has a number of advantages over analog modulation (Goldsmith 2005, Proakis 2000). Firstly, by adopting high-order modulation techniques such as MQAM, digital modulation offers a high data rate as compared to analog modulation with the same signal bandwidth. Secondly, in systems of digital modulations, the ISI problem is mitigated by using powerful channel coding,

https://doi.org/10.1515/9783110751437-004

equalization or multicarrier techniques. Thirdly, spread spectrum techniques can also be used in the system of digital modulation to remove or combine multipath, anti-interference and detect multiple users simultaneously. Finally, encryption can be easily combined with digital modulation to provide a higher level of security and privacy. For the reasons above, the current wireless systems are almost all digital modulated systems.

4.1 Transmitter and receiver diagrams

Figure 4.2 shows a simplified transmitter and Fig. 4.3 shows the corresponding receiver.

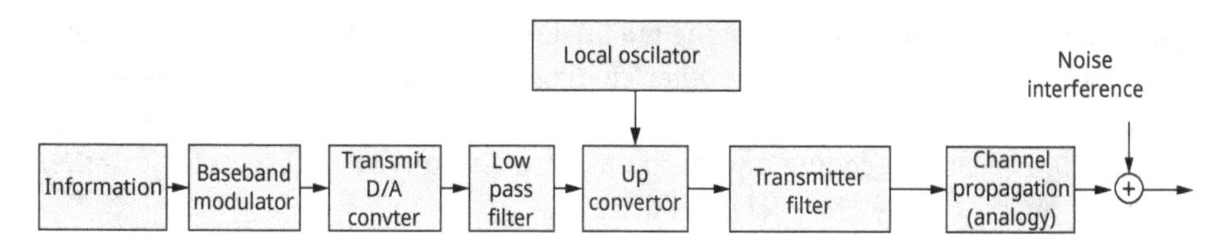

Fig. 4.2: Simplified block diagram of a digital transmitter.

In Fig. 4.2, the information may be data information, or it also can be converted from an analogy information source. If it is from an analogy information source, it must be first converted to a digital signal by A/D convertor, and then it might be followed by a source encoder and channel encoder. In multiple-user communication systems, signaling and multiplexing components should be used. This diagram focuses on the modulation part. It consists of the following five components: baseband modulation, transmission A/D convertor, low pass filter, up converter, local oscillator and transmit filter. This part will be discussed in detail in this chapter.

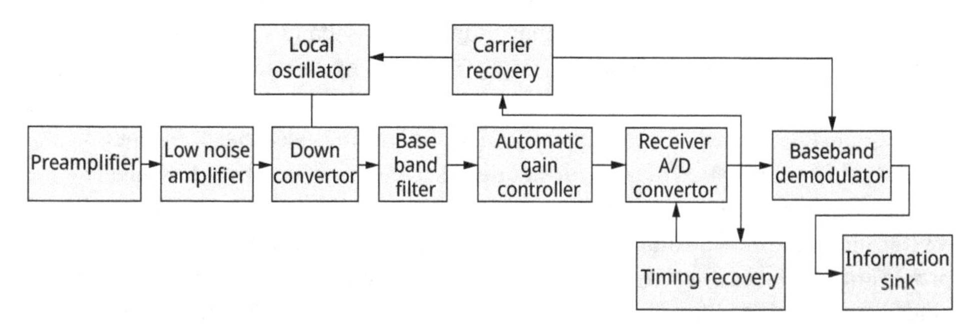

Fig. 4.3: Block diagram of a digital receiver.

Similarly, Fig. 4.3 also focuses on the demodulation component. We assume a receiver filter has already eliminated the out-of-band emissions in the RF domain before the preamplifier, and then the signal is amplified by a low noise amplifier. The local oscillator (LO) which is generated and tracked by a carrier frequency recovery component provides an unmodulated sinusoidal signal, to convert the radio frequency signal into a baseband signal. An automatic gain controller generally is an optional component that is determined by the dynamic range of the received signal. After being sampled, which is done by the receiver A/D convertor, the received signal is put into the baseband demodulator. Of course, the sampling points of the receiver A/D convertor are provided by timing recovery which generally is a tracking loop. If the communication system is a multi-user one, de-multiplexed will be followed. In practice, channel decoding might follow the demodulation. If the information is not data by analogy, a source decoder and D/A convertor are needed to obtain the analogy signal.

4.2 Signal space representation

In Fig. 4.2, the task of baseband modulation is to encode a bit stream of finite length into a pulse sequence. After being D/A converted, low pass filtered and up converted, the pulse sequence will be transformed into one of several possible transmitted signals. At the receiver, the received signal will be decoded in the set of possible transmitted signals and the one which is "closest" to the received signal will be decoded as the transmitted signal.

In order to find the distance between the received signal and the possible transmitted signals, we represent the signals as projections onto a set of basis functions. In such a way, we set up a one-to-one mapping between the set of transmitted signals and their vector representations, which is a finite-dimensional vector space.

4.2.1 Signal space and constellation

Assume that there are $M = 2^K$ possible sequences of K bits, and each $K = \log_2 M$ bits are to be sent in T seconds, thus the data rate is $R = K/T = \log_2 M/T$ bps. The M possible sequence $M = \{m_1, m_2, \ldots, m_M\}$ is the set of all messages or symbols, and each of them is a K bits symbol, $\{b_1, b_2, \ldots, b_K\} \in M$ and each message has a probability to be selected to be sent. Now, let's discuss how to map these bits sequences to analogy signals.

As we know, suppose $S = (s_1(t), \ldots, s_M(t))$ is any set of M real energy signals defined on $[0, T)$, by certain procedures like the Gram–Schmidt orthogonalization, the signals can be represented by a linear combination of $N \leq M$ real orthonormal basis functions $\{\varphi_1(t), \ldots, \varphi_N(t))\}$. In this way, the set S is said to be spanned by these basis functions. Therefore, we can represent each $s_i(t) \in S$ as

$$s_i(t) = \sum_{j=1}^{N} s_{ij}\phi_j(t) \qquad 0 \le t < T \tag{4.1}$$

where T is the symbol time and

$$s_{ij} = \int_0^T s_i(t)\phi_j(t)dt \tag{4.2}$$

is a real coefficient representing the projection of $s_i(t)$ onto the basis function, apparently $(s_{i1}, s_{i2}, \ldots, s_{iN})$ make up a vector. And the basis functions satisfy the orthogonal condition

$$\int_0^T \phi_i(t)\phi_j(t)dt = \begin{cases} 1 & i = j \\ 0 & i \ne j \end{cases} \tag{4.3}$$

For linear passband modulations, the basis set consists of the sine and cosine functions:

$$\phi_1(t) = \sqrt{\frac{2}{T}}\cos(2\pi f_c t) \tag{4.4}$$

And

$$\phi_2(t) = \sqrt{\frac{2}{T}}\sin(2\pi f_c t) \tag{4.5}$$

With the basis function above, eq. (4.1) can be rewrite as

$$s_i(t) = s_{i1}\sqrt{\frac{2}{T}}\cos(2\pi f_c t) + s_{i2}\sqrt{\frac{2}{T}}\sin(2\pi f_c t) \tag{4.6}$$

The vector $\mathbf{s}_i = (s_{i1}, \ldots, s_{iN})$ is called the *signal constellation point* corresponding to the signal $s_i(t)$. We call all constellation points $\{\mathbf{s}_1, \ldots, \mathbf{s}_M\}$ the *signal constellation.*

So far, we have set up a mapping relation between the modulation signals and the signal constellation points. In this way, we change the distance calculation of the distance of two real signals, which are the received signal and a transmitted modulated signal, to the calculation of two vectors, which are the received vector and a signal constellation point. In such a way, the fundamental of modulation is to convert the message to be transmitted into the continuous transmitted signal through the corresponding constellation point. The fundamental of demodulation is to convert the received continuous signal into a point (vector) in the signal space and find the nearest constellation point to this vector. The result constellation point is then translated to the decoded information message.

A two-dimensional signal space can be found in Fig. 4.4. In the space, a constellation point $\boldsymbol{s}_i$ belongs to R^2, and the ith axis of R^2 corresponds to the basis function $\phi_i(t)$, $i = 1, 2$. Actually, $\phi_1(t)$ and $\phi_2(t)$ are the functions in eqs. (4.4) and (4.5).

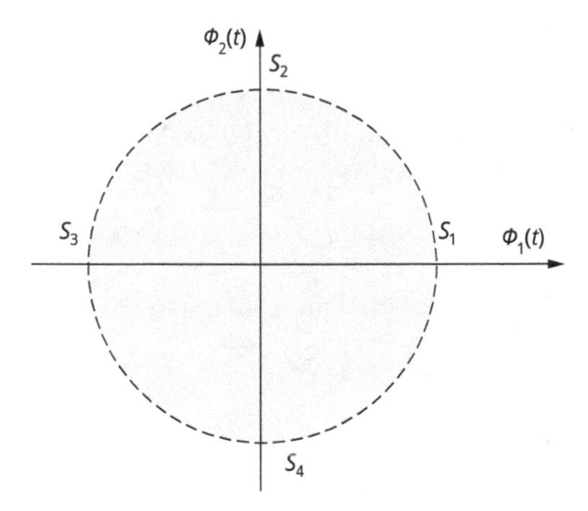

Fig. 4.4: Two-dimensional signal space representation.

Linear modulation techniques use two-dimensional signal space representations, where the two dimensions correspond to the in-phase and quadrature basis functions. We'll discuss such kinds of modulations like MPSK and MQAM in this chapter.

Suppose there are $M = 2^K$ possible sequences of K bits, a bit sequence of length K comprises a message $m_i = \{\boldsymbol{b}_1, \ldots, \boldsymbol{b}_K\} \in M$, and $M = \{m_1, \ldots, m_M\}$ is the set of messages. Generally, each message has a probability of being sent in the communication system. Let p_i is the probability of m_i being selected for the transmission and the message m_i is the symbol sent over to the channel during the time interval $[0, T)$. Since the actual channel is analog, the message is first transformed into an analog signal through eq. (4.1) for channel transmission. Thus, we mapped each signal message $m_i \in M$ to a unique analog signal $s_i(t) \in S = \{s_1(t), \ldots, s_M(t)\}$, where $s_i(t)$ is signal defined on a symbol duration $[0, T)$ with an energy of

$$s_i(t) = s_{i1}\sqrt{\frac{2}{T}}\cos(2\pi f_c t) + s_{i2}\sqrt{\frac{2}{T}}\sin(2\pi f_c t) \tag{4.7}$$

For the AWGN channel, the probability of received vector $\boldsymbol{r}$ conditioned on the transmitted constellation point $\mathbf{s}_i$ is considered. The received vector element r_j is a Gauss-distributed random variable, and r_j and r_k, are independent for $k \neq j$. The mean of r_j is s_{ij} and the variance is $N_0/2$. The conditional distribution of $\boldsymbol{r}$ is thus given as follows:

$$p(r|s_i sent) = \prod_{j=1}^{N} p(r_j|m_i) = \frac{1}{(\pi N_0)^{N/2}} \exp\left[-\frac{1}{N_0}\sum_{j=1}^{N}(r_j - s_{ij})^2\right] \tag{4.8}$$

4.2.2 Decision regions and the maximum likelihood decision criterion

In the receiver design, we try to minimize the demodulation error P_e, i.e. to maximize

$$1 - P_e = p(\hat{m} = m_i|\boldsymbol{r}) \tag{4.9}$$

and this is equivalent to maximizing $p(\hat{m} = mi|r(\mathrm{t}))$. Because there is a one-to-one mapping between messages and signal constellation points, this equals to maximizing $p(\mathbf{s}_i$ sent$|r(t))$. It is same to find the desired vector $\boldsymbol{r}$ that satisfies

$$p(\mathbf{s}_i sent|\boldsymbol{r})\rangle\ p(\mathbf{s}_j sent|\boldsymbol{r}) \quad \forall j \neq i \tag{4.10}$$

The *decisions regions* ($Z_1, \ldots, Z_M$) is defined as the subsets of the signal space R^N by

$$Z_i = (\boldsymbol{r}: p(\mathbf{s}_i sent|\boldsymbol{r}) > p(\mathbf{s}_j sent|\boldsymbol{r}) \forall j \neq i) \tag{4.11}$$

The inequality above indicates these regions are not overlapping and the points that make $p(\mathbf{s}_i sent|\boldsymbol{r}) = p(\mathbf{s}_j sent|\boldsymbol{r})$ do not belong to any decision regions. Since the whole signal space has been divided into different decision regions, for a given received vector $\boldsymbol{r} \in Z_i$, the output of the optimal receiver is an estimated message $\hat{m} = m_i$.

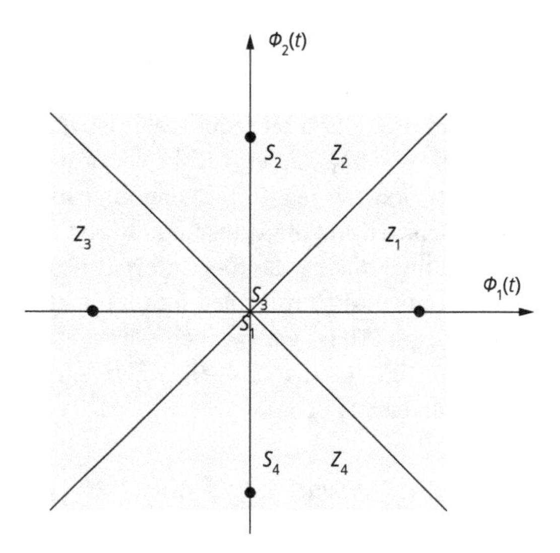

Fig. 4.5: Decision regions for a two-dimensional signal space and four constellation points.

The processing of receiver is computing the received vector $\boldsymbol{r}$ from $r(t)$, it is equivalent to find the decision region Z_i contains $\boldsymbol{r}$, and this decision region Z_i corresponds to the message m_i. Figure 4.5 gives a two-dimensional signal space, where the four decision regions $Z_1, \ldots, Z_4$ correspond to four constellations $\mathbf{s}_1, \ldots, \mathbf{s}_4$. The received vector $\boldsymbol{r}$ lies in region Z_1 means the message m_1 is the desired message given received vector $\boldsymbol{r}$.

By Bayes rule,

$$p(\boldsymbol{s}_i|\boldsymbol{r}) = \frac{p(\boldsymbol{r}|\boldsymbol{s}_i)p(\boldsymbol{s}_i)}{p(\boldsymbol{r})} \tag{4.12}$$

Minimizing the symbol error probability corresponds to find the constellation $\boldsymbol{s}_i$ that maximizes $p(\boldsymbol{s}_i|\boldsymbol{r})$. Since $p(\boldsymbol{r})$ is not a function of $\boldsymbol{s}_i$, and we assume the messages are equally likely

$$\begin{aligned} \arg\max_{s_i} \tfrac{p(\boldsymbol{r}|\boldsymbol{s}_i)p(\boldsymbol{s}_i)}{p(\boldsymbol{r})} &= \arg\max_{s_i} p(\boldsymbol{r}|\boldsymbol{s}_i)p(\boldsymbol{s}_i),\ i=1,2,\ldots,M \\ &= \arg\max_{s_i} p(\boldsymbol{r}|\boldsymbol{s}_i),\ i=1,2,\ldots,M \end{aligned} \tag{4.13}$$

Let's define the likehood function associated to the receiver as

$$L(\boldsymbol{s}_i) = p(\boldsymbol{r}|\boldsymbol{s}_i) \tag{4.14}$$

And the corresponding log likelihood function is

$$l(\boldsymbol{s}_i) = \log(L(\boldsymbol{s}_i)) \tag{4.15}$$

Since the log function is an increasing function with its argument, maximizing eq. (4.15) are equivalent to maximizing eq. (4.14). Substitute eqs. (4.8) into (4.14) then to (4.15), we obtain

$$l(\boldsymbol{s}_i) = -\frac{1}{N_0}\sum_{j=1}^{N}(\boldsymbol{r}_j - \boldsymbol{s}_{ij})^2 = -\frac{1}{N_0}||\boldsymbol{r} - \boldsymbol{s}_i|| \tag{4.16}$$

From eq. (4.16), we see that the value of the log likelihood function $l(\boldsymbol{s}_i)$ is related to the distance between the received vector $\boldsymbol{r}$ and the constellation point $\boldsymbol{s}_i$.

The result constellation point $\boldsymbol{s}_i$ is the decision region Z_i contains the vector $\boldsymbol{r}$, where Z_i is

$$Z_i = (\boldsymbol{r}: ||\boldsymbol{r} - \boldsymbol{s}_i|| < ||\boldsymbol{r} - \boldsymbol{s}_j||,\ j \neq i, i = 1, 2, \ldots, M) \tag{4.17}$$

If an error occurs, there must be another constellation point $\boldsymbol{s}_k$ that makes the distance between $\boldsymbol{r}$ and $\boldsymbol{s}_i$ is great than that of $\boldsymbol{r}$ and $\boldsymbol{s}_k$, where $k \neq i$, i.e.

$$||\boldsymbol{r} - \boldsymbol{s}_k|| < ||\boldsymbol{r} - \boldsymbol{s}_i||,\ k \neq i \tag{4.18}$$

Since the channel is additive white Gaussian noise channel, eq. (4.18) implies the noise projected on the vector $\boldsymbol{s}_k$–$\boldsymbol{s}_i$ is greater than half of distance between $\boldsymbol{s}_k$ and $\boldsymbol{s}_i$. Assume projecting $\boldsymbol{n}$ onto a one-dimensional line yields a one-dimensional Gaussian random variable n with a mean and variance $N_0/2$,

$$p(||\boldsymbol{r}-\boldsymbol{s}_k||<||\boldsymbol{r}-\boldsymbol{s}_j||)=p(n>d_{ik}/2)=\int_{d_{ik}}^{\infty}\frac{1}{\sqrt{\pi N_0}}\exp\left(\frac{-v^2}{N_0}\right)dv=Q\left(\frac{d_{ik}}{\sqrt{2N_0}}\right) \tag{4.19}$$

where $Q(z)$ is the so-called Q function defined as the probability when x is bigger than z for x is a Gaussian random variable with mean 0 and variance 1:

$$Q(z)=p(x>z)=\int_{z}^{\infty}\frac{1}{\sqrt{2\pi}}\exp\left[\frac{-x^2}{2}\right]dx \tag{4.20}$$

The symbol error probability P_e is sometimes represented by the nearest neighbor approximation, which is defined as the multiplication of the probability of error associated with constellations at the minimum distance d_{min} and the number of neighbors with the minimum distance M_{dmin}

$$P_e \approx M_{min}Q\left(\frac{d_{min}}{\sqrt{2N_0}}\right) \tag{4.21}$$

Recall that a message consists of $\log_2 M$ bits and we always evaluate the system performance by the bit error probability (BEP) or bit error rate (BER) P_b. In practice, the mapping of the M possible bit sequences to messages m_i, $i=1,\ldots,M$ should be well designed so that one symbol error is caused by the most likely way to make an error, which is associated with an adjacent decision region, corresponds to only one-bit error, which is the case mistaken by an adjacent decision region, corresponds to only one-bit error. With such a mapping, we can make the approximation

$$P_b \approx \frac{P_e}{\log_2 M} \tag{4.22}$$

4.3 Modulation principles and formats

In this chapter the equivalent baseband representation in eq. (2.3) for the description of modulation signals is used.

$$s(t)=\mathrm{Re}\{u(t)\exp[j2\pi f_c t])\} \tag{4.23}$$

We first discuss the simplest format of modulation: *Pulse Amplitude Modulation,* then we will discuss other important modulation formats.

4.3.1 Pulse amplitude modulation (PAM)

Pulse amplitude modulation is the simplest form of linear modulation, which has a one-dimensional signal space and has only one basic function. For MPAM, all of the information is encoded into the signal amplitude A_i. The equivalent baseband representation for the transmitted signal is given by

$$u(t) = \sum_{i=-\infty}^{\infty} A_i g(t - iT_s) \tag{4.24}$$

where $A_i = (2i - 1 - M)\, d$, $i = 1, 2, \ldots, M$ defines the one-dimensional signal constellation and $g(t)$ is the pulse shape filter.

Substitute eq. (4.24) into eq. (4.23), we obtain

$$s(t) = \sum_{i=-\infty}^{\infty} A_i g(t - iT_s) \cos(2\pi f_c t) \tag{4.25}$$

For a signal in a specific symbol time, $s(t)$ changes to $s_i(t)$

$$s_i(t) = A_i g(t) \cos(2\pi f_c t) \tag{4.26}$$

The difference in various PAM formats lines in how the data bits are mapped to the constellation points and then to the modulation coefficients. The minimum distance between two adjacent constellation points is $d_{min} = \min_{i,j}|A_i - A_j| = 2d$. The information bits are conveyed by M different values of the amplitude of the transmitted signal and $\log_2 M = K$ bits per symbol time T_s is represented by each pulse. Apparently, the energy of different constellation points is different.

$$E_{s_i} = \int_0^{T_S} s_i^2(t)dt = \int_0^{T_S} A_i^2 g^2(t) \cos^2(2\pi f_c t)dt = A_i^2 \tag{4.27}$$

We generally assume that basis pulses are normalized to unit average power, say

$$\int_0^{T_S} |g(t)|^2 dt = 1 \tag{4.28}$$

The constellation mapping can be done in Gray coding, which can be shown in Fig. 4.6.

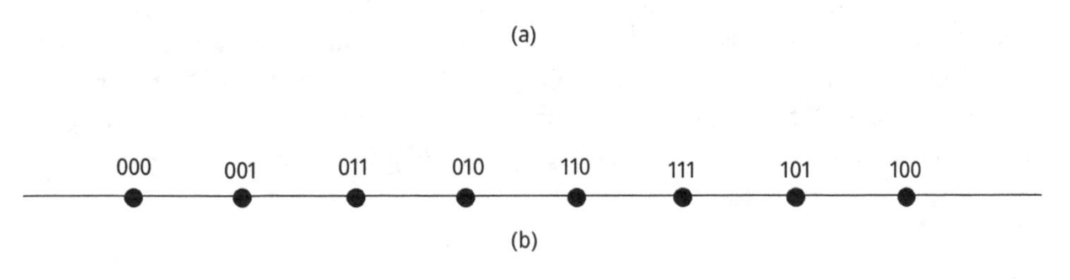

Fig. 4.6: Gray coding for 4PAM (a) and 8PAM (b).

Equation (4.28) gives the pulse-shaping power constraint; however, a specific pulse shape is required so that no inter-symbol interference will be introduced. The pulse shaping will be discussed later.

4.3.2 Multiple phase shift keying (MPSK)

Different from MPAM where information is conveyed by the amplitude of the modulated signal, in multiple phase shift keying (MPSK), the information is conveyed by in the phases of the modulated signal.

4.3.2.1 Binary phase shift keying (BPSK)

BPSK modulation is the simplest PSK modulation: there are totally two symbols, the bit 0 and bit 1, which are conveyed by the phase 0 and π separately.

$$u(t) = \sum_{k=-\infty}^{\infty} Ag(t-kT_s)\exp(j2\pi(i_k-1)/M) \tag{4.29}$$

where $M = 2$ for BPSK modulation.

Substitute eq. (4.29) into (4.23), we obtain

$$\begin{aligned} s(t) &= \mathrm{Re}\left\{\sum_{k=-\infty}^{\infty} Ag(t-kT_s)\exp(j2\pi(i-1)/M)\exp[j2\pi f_c t])\right\} \\ &= \sum_{k=-\infty}^{\infty} Ag(t-kT_s)\cos(2\pi f_c t + 2\pi(i_k-1)/M)) \end{aligned} \tag{4.30}$$

where $M = 2$.

For a specific symbol time,

$$s_1(t) = \cos(2\pi f_c t) \qquad \text{for } \mathrm{m}_1 \text{ equals bit 1} \tag{4.31}$$

$$s_2(t) = -A\cos(2\pi f_c t) \qquad \text{for } \mathrm{m}_2 \text{ equals bit 0} \tag{4.32}$$

Figure 4.7(a) shows the symbol pulse of the symbol sequence and Fig. 4.7(b) shows the corresponding signal.

Figure 4.8 shows the signal space of BPSK. Recall that the constellation points A and $-A$ related to the energy of a bit in BPSK. For rectangular pulse shaping of BPSK, $E_b = A^2$, so $A = \sqrt{E_b}$. For the one-dimensional signal space, there is only one basis function. So the modulation and demodulation have only one branch. Figure 4.9 gives modulation and the corresponding coherent demodulation of BPSK.

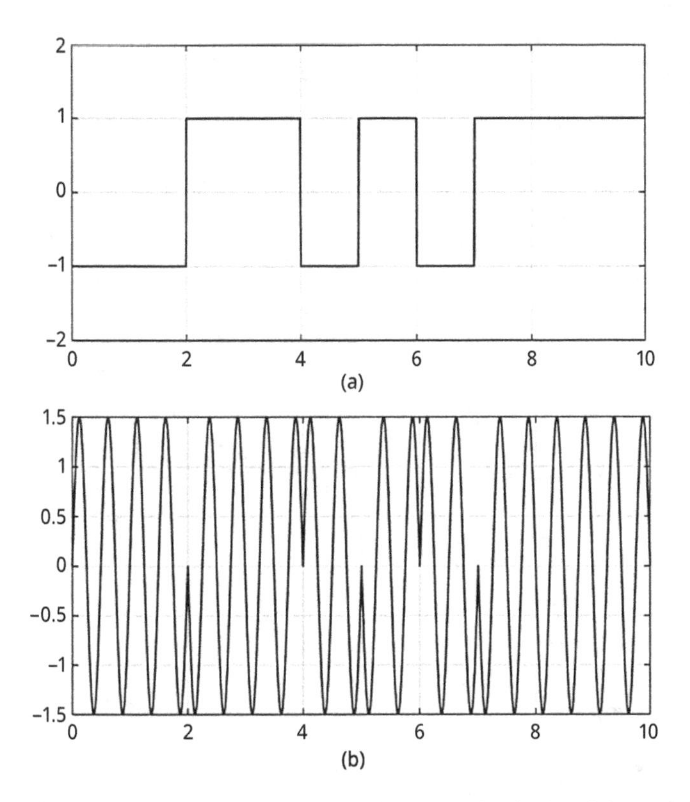

Fig. 4.7: Binary phase shift keying signal: (a) Normalized pulse of the symbol sequence (b) The corresponding BPSK signal.

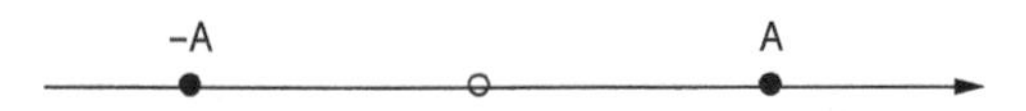

Fig. 4.8: Signal space for BPSK.

4.3.2.2 Quadrature-phase shift keying (QPSK) and multiple phase shift keying (MPSK)

It can be considered that quadrature-phase shift keying (QPSK) are two branches of BPSK. Figure 4.10 gives the signal constellation. We can find that there is a BPSK signal space in each dimension. The Gray coding is also shown in the signal space.

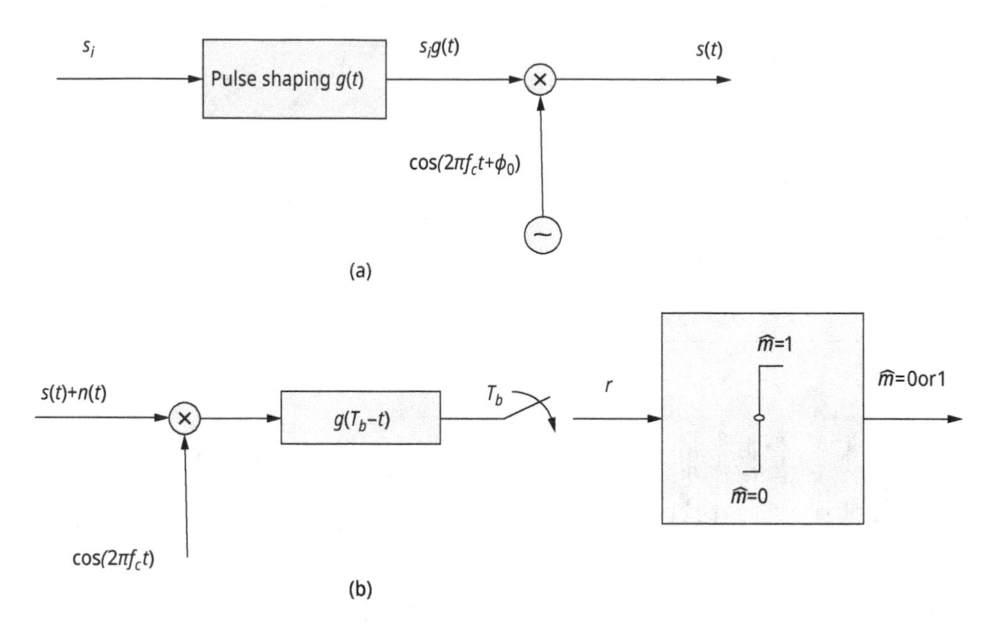

Fig. 4.9: Modulation diagram of BPSK (a) and its coherent demodulation (b).

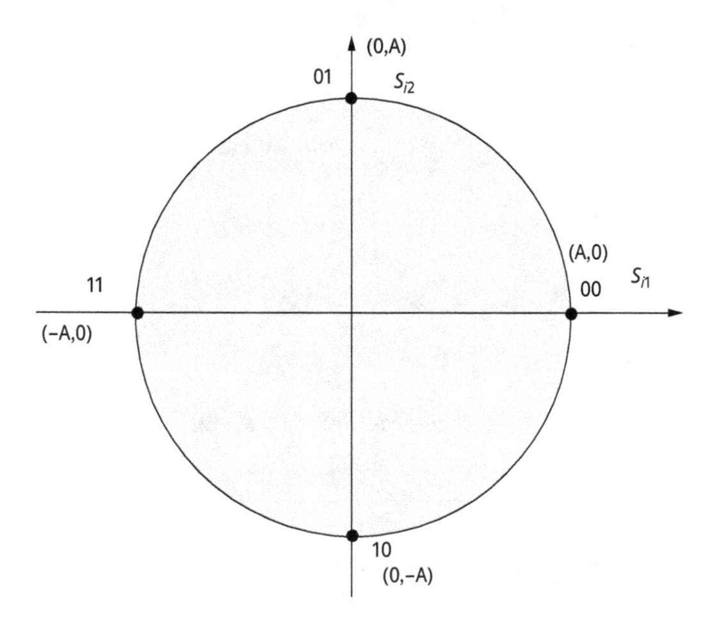

Fig. 4.10: Signal space for QPSK and the corresponding Gray coding.

The original data stream is split into two streams, b_{1i} and b_{2i},

$$\left.\begin{aligned} b_{1i} &= b_{2i} \\ b_{2i} &= b_{2i+1} \end{aligned}\right\} \tag{4.33}$$

each of which has a data rate that is half of the original data stream:

$$R_s = 1/T_s = R_B/2 = 1/(2T_B) \tag{4.34}$$

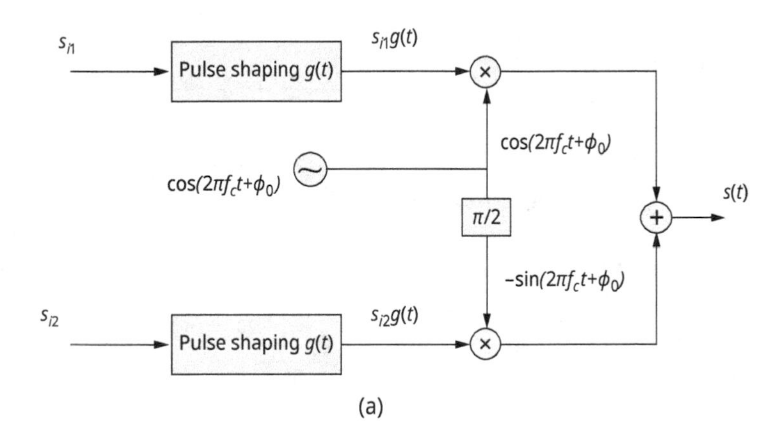

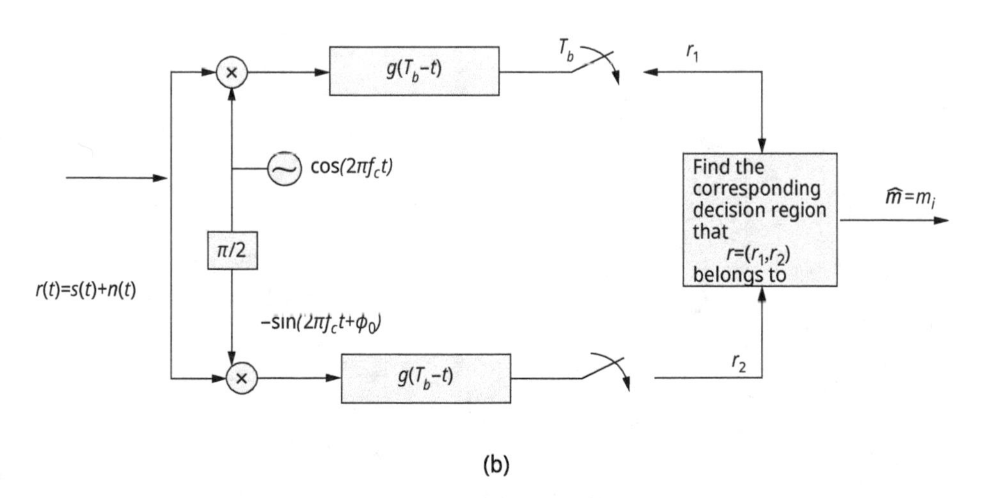

Fig. 4.11: Modulation of MPSK/MQAM (a) and it coherent demodulation (b).

The modulation and coherent demodulation of QPSK are shown in Fig. 4.11. In this Fig. 4.11(a), the two data streams are modulated in the in-phase and quadrature components separately, each of them is a BPSK. It should be mentioned that Fig. 4.11 is not only the modulation and coherent demodulation of QPSK, but also a general diagram of modulation and coherent demodulation of amplitude and phase modulations. And of course, for QPSK, the modulation and demodulation can be simplified to Fig. 4.12. In the general case, the data stream cannot be simply split into two data streams. For each symbol, it is mapped the corresponding constellation point, then the in-phase and quadrature components can be found, which can be modulated by the two branches. The in-phase or quadrature components may not have corresponding split streams. BPSK and MPAM can be seen as a special case of it, in which only the upper in-phase branch is kept and the quadrature branch is omitted.

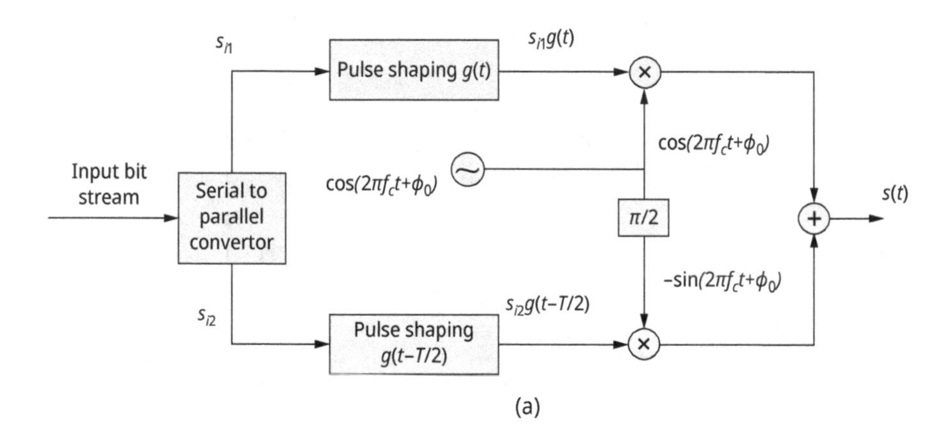

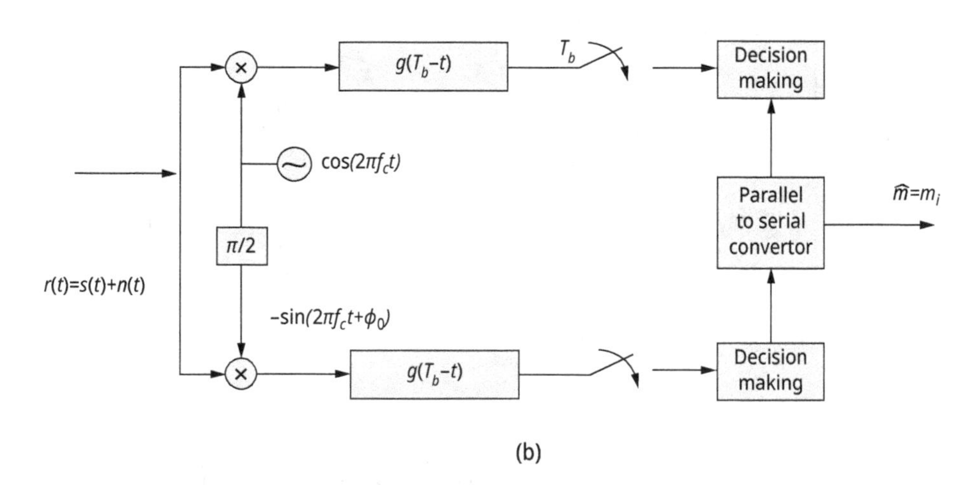

Fig. 4.12: Modulation of QPSK (a) and it coherent demodulation (b).

For a general MPSK, $u(t)$ is exactly the same as (4.29) and $s(t)$ is same as (4.30), except that M is not limit to 2, but $M = 2^k$, where k is the number of bits in a symbol. For a specific symbol time, the modulated signal becomes

$$\begin{aligned} s_i(t) &= \mathrm{Re}\{Ag(t)\exp(j2\pi(i-1)/M)\exp[j2\pi f_c t])\} \\ &= Ag(t)\cos(2\pi f_c t + 2\pi(i-1)/M) \\ &= Ag(t)\cos(2\pi(i-1)/M)\cos 2\pi f_c t - Ag(t)\sin(2\pi(i-1)/M)\sin 2\pi f_c t \end{aligned} \tag{4.35}$$

Since both the in-phase and the quadrature-phase components of QPSK are exploited for the transmission of information, the spectral efficiency of QPSK is twice the efficiency of BPSK, and we can transmit twice the data rate as that of BPSK with the same bandwidth.

4.3.2.3 Quadrature offset QPSK

In the signal space representation, each symbol $\mathbf{s}_i = (s_{i1}, s_{i2})$ of QPSK lies in one of the four quadrants. At each of the new symbol time, a 180-degree phase transition may occur due to the constellation point transition, and distortion will be introduced by this phase transition and quick amplitude variation when the signal pass through nonlinear amplifiers and filters. Quadrature offset QPSK is another kind of QPSK in which the in-phase and the quadrature-phase components occur at different time instants. The modulation and coherent demodulation of OQPSK is shown in Fig. 4.13.

4.3.2.4 π/4-QPSK and π/4-DQPSK

$\pi/4$-QPSK ($\pi/4$ differential quadrature-phase shift keying) is another technique to mitigate the amplitude fluctuations of a 180 degree phase shift. It reduces the phase transition from 180 degrees of QPSK to 135 degrees. Comparing with the phase transition of 90 degrees of OQPSK, the spectral properties under nonlinear amplification of $\pi/4$-QPSK is not as good as that of OQPSK. However, the $\pi/4$-QPSK can be done after the information bits be differentially encoded, so that we do not need a coherent phase reference to demodulate the information. This kind of $\pi/4$-QPSK is called $\pi/4$-DQPSK.

Figure 4.14(a) and (b) show two QPSK constellations that are shifted by $\pi/4$ with respect to each other. All the possible state transition is shown in Fig. 4.14(c). Switching between the constellation (a) and (b) means that there is at least a phase transition of an integer multiple of $\pi/4$ between successive symbols and time recovery and synchronization can be done according to this transition.

The differential coding of the information can be described in Tab. 4.1, which lists the carrier phase shift corresponding to various input bit pairs.

Figure 4.15 gives the modulation of π/4-QPSK, where the signal-mapping module works as follows:

$$\begin{aligned} I_k &= \cos\theta_k = I_{k-1}\cos\phi_k - Q_{k-1}\sin\phi_k \\ Q_k &= \sin\theta_k = I_{k-1}\sin\phi_k + Q_{k-1}\cos\phi_k \end{aligned} \tag{4.36}$$

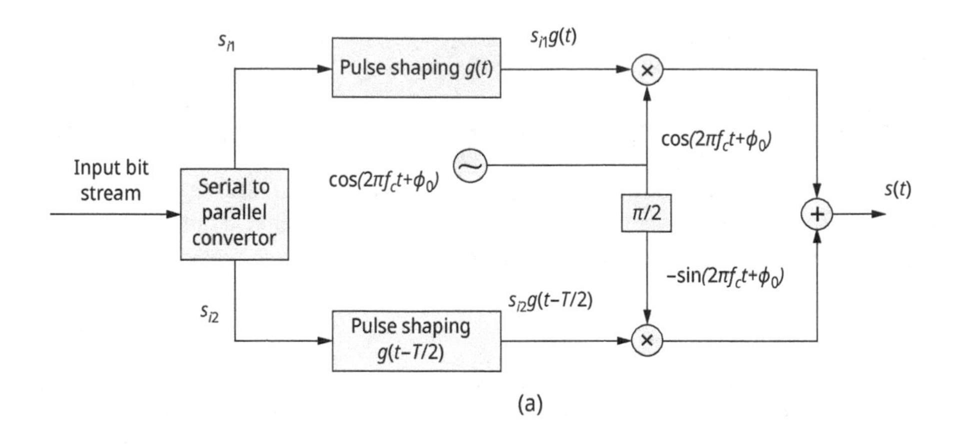

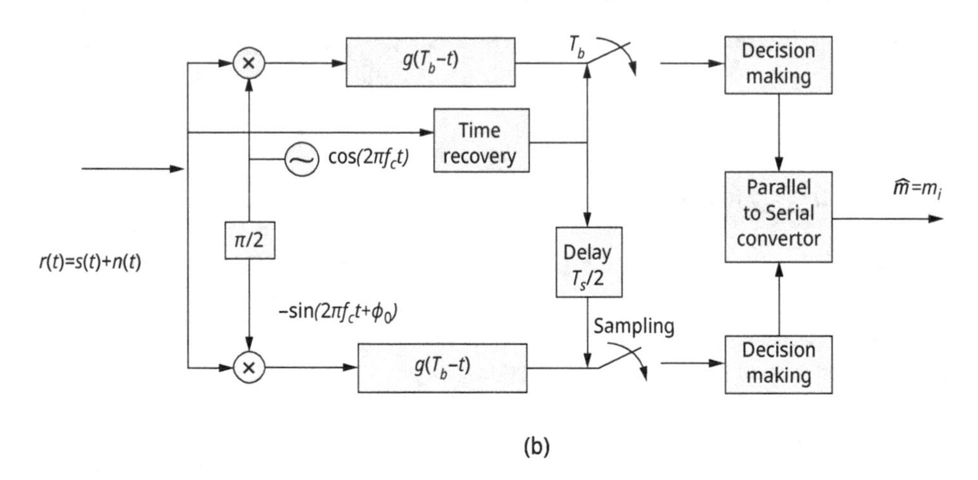

Fig. 4.13: Modulation of OQPSK(a) and its coherent demodulation (b).

where

$$\theta_k = \theta_{k-1} + \phi_k \tag{4.37}$$

and ϕ_k is given in Tab. 4.1.

The $\pi/4$-DQPSK modulated signal can be coherently detected or differentially detected. The BER performance of the latter case is about 3 dB worse than that of the former case, which has the same error performance as OPSK. However, the differential demodulation is easy to implement in hardware and offers a lower error floor in fast Rayleigh fading channels. Figure 4.16 gives the differential demodulation diagram.

Assume $\phi_k = \arctan(Q_k/I_k)$ is the phase of the carrier due to the kth data bit, the output w_k from the matched filter in the in-phase branches of the demodulator can be expressed as

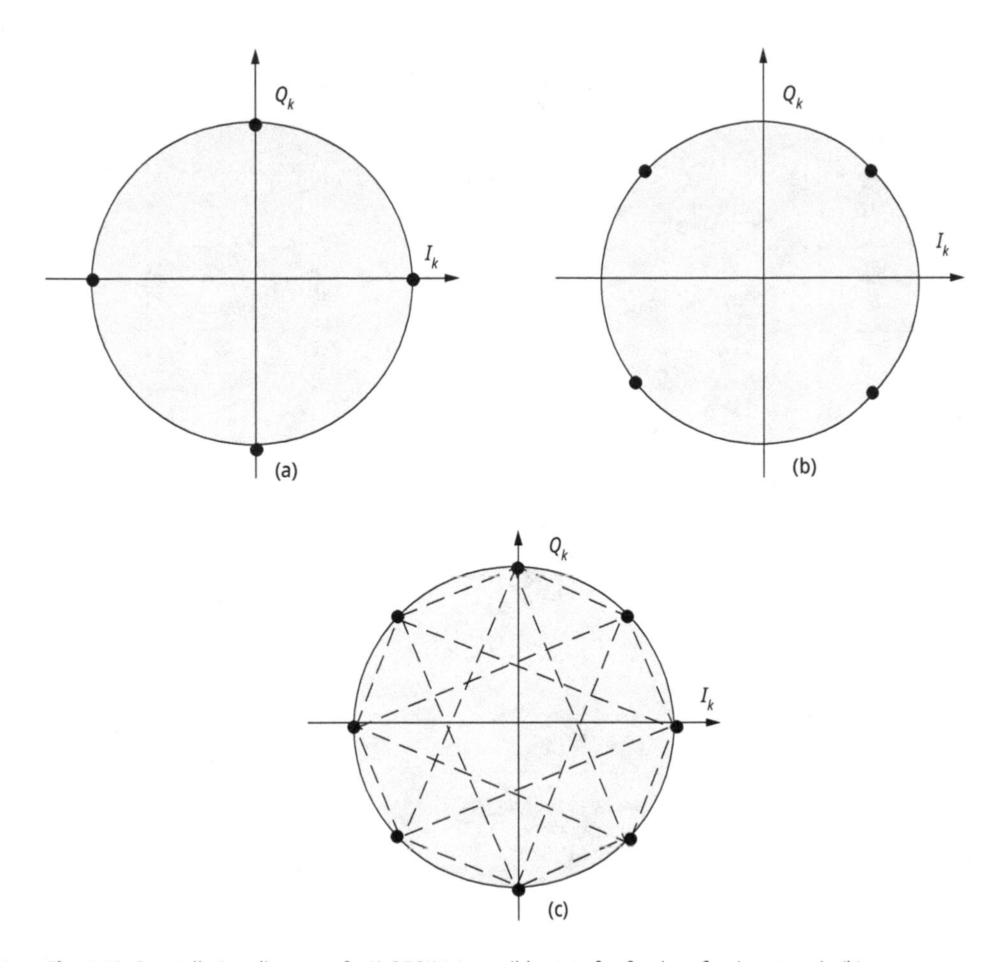

Fig. 4.14: Constellation diagram of $\pi/4$-QPSK (a) possible state for θ_k when θ_{k-1} is a state in (b); (b) possible state for θ_k when θ_{k-1} is a state in (a); all possible states transition.

Tab. 4.1: Carrier phase shift corresponding to various input bit pairs.

Information bit $m_{Ik}m_{Ok}$	Phase shift ϕ_k
11	$\pi/4$
01	$3\pi/4$
00	$-3\pi/4$
10	$-\pi/4$

$$w_k = \cos(\phi_k - \gamma) \tag{4.38}$$

Similarly, the output z_k from the matched filter in the quadrature branch is

$$z_k = \sin(\phi_k - \gamma) \tag{4.39}$$

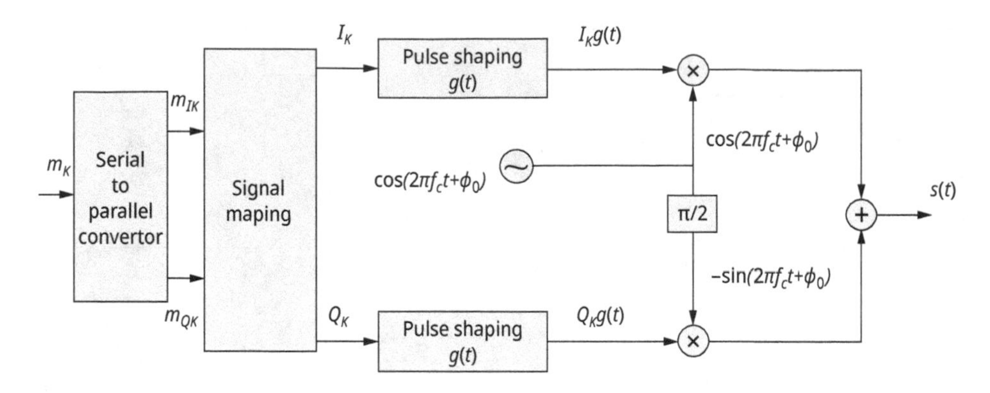

Fig. 4.15: Modulation of π/4-DQPSK.

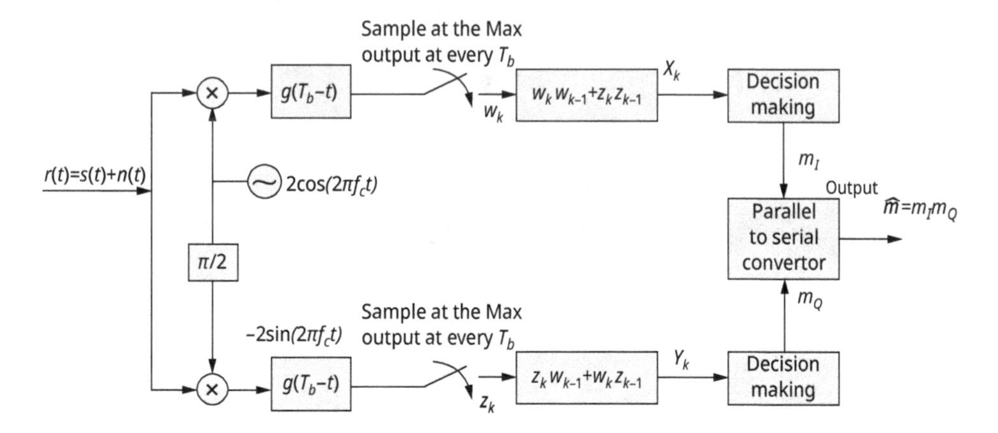

Fig. 4.16: Demodulation of π/4-DQPSK.

In the equations above, γ is a phase shift caused by noise, propagation delay and interference which is assumed to be essentially constant because its variation is slow comparing to the phase component ϕ_k. Put w_k and z_k into a differential decoder, we get:

$$\begin{aligned} x_k &= w_k w_{k-1} + z_k z_{k-1} \\ &= \cos(\phi_k - \gamma)\cos(\phi_{k-1} - \gamma) + \sin(\phi_k - \gamma)\sin(\phi_{k-1} - \gamma) \\ &= \cos(\phi_k - \phi_{k-1}) \end{aligned} \tag{4.40}$$

$$\begin{aligned} y_k &= z_k w_{k-1} - w_k z_{k-1} \\ &= \sin(\phi_k - \gamma)\cos(\phi_{k-1} - \gamma) + \cos(\phi_k - \gamma)\sin(\phi_{k-1} - \gamma) \\ &= \sin(\phi_k - \phi_{k-1}) \end{aligned} \tag{4.41}$$

By applying Tab. 4.1 to the decision making component, one can make a decision that

$$m_I = 1 \quad \text{if} \quad x_k > 0 \quad \text{or} \quad S_I = 0 \quad \text{if} \quad x_k < 0 \tag{4.42}$$

$$m_Q = 1 \quad \text{if} \quad y_k > 0 \quad \text{or} \quad S_Q = 1 \quad \text{if} \quad y_k < 0 \tag{4.43}$$

where m_I and m_Q are the demodulated bits in the in in-phase and quadrature branches, respectively.

4.3.3 Quadrature amplitude modulation (MQAM)

Among the linear modulation forms, MQAM is the most spectral efficient, where the information bits are conveyed in both the amplitude and phase of the modulated signal. As we have discussed above, there is only one degree of freedom to encode the information bits (amplitude or phase) in MPAM and MPSK. However, there are two degrees of freedom in MQAM. That is why it has a high spectrally efficiency.

In eq. (4.23), the $u(t)$ change to

$$u(t) = \sum_{k=-\infty}^{\infty} A_i g(t - kT_s) \exp(j\theta_i) \tag{4.44}$$

Substitute eq. (4.44) into (4.23) yields

$$\begin{aligned} s(t) &= \text{Re}\left\{ \sum_{k=-\infty}^{\infty} A_i g(t - kT_s) \exp(j\theta_i) \exp[j2\pi f_c t]) \right\} \\ &= \sum_{k=-\infty}^{\infty} A_i g(t - kT_s) \cos(2\pi f_c t + 2\pi(i_k - 1)/M)) \end{aligned} \tag{4.45}$$

For a specific symbol time T_i,

$$\begin{aligned} s_i(t) &= R\{A_i e^{j\theta_i} g(t) e^{j2\pi f_c t}\} \\ &= A_i \cos(\theta_i) g(t) \cos(2\pi f_c t) - A_i \sin(\theta_i) g(t) \sin(2\pi f_c t) \end{aligned} \tag{4.46}$$

For MQAM with square signal constellations, s_{i1} and s_{i2} take values on $(2i - 1 - L)\, d$, $i = 1, 2, \ldots L = 2l$, it can be easily found that the minimum distance between signal points is $d_{\min} = 2d$, which is exactly the same to that of MPAM. Actually, a square constellation MQAM with a constellation size of L^2 and MPAM modulation with a constellation size of L on each of the in-phase and quadrature signal branches can be regarded as equivalent. Figure 4.17 shows the 16QAM square constellation and Fig. 4.18 gives the corresponding decision region.

The modulation and the coherent demodulation principle of MQAM are shown in Fig. 4.12, which is exactly same as that of MPSK.

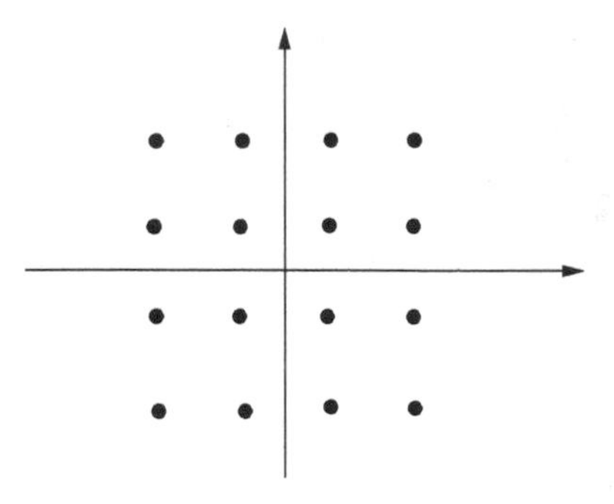

Fig. 4.17: 16QAM constellation.

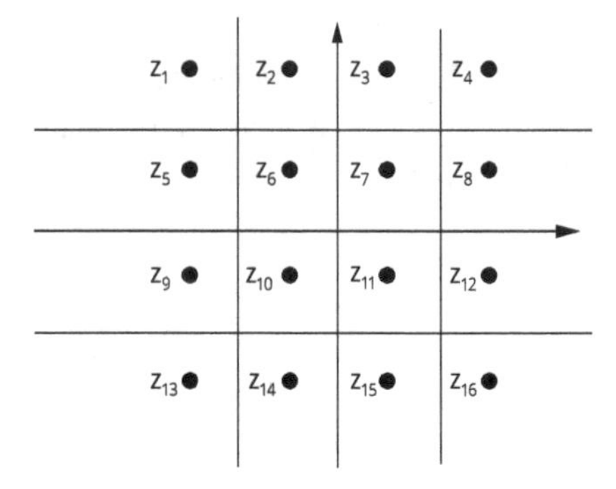

Fig. 4.18: Decision regions for square constellation 16QAM.

4.3.4 Frequency modulation

4.3.4.1 Multiple frequency shift keying (MFSK)

Frequency modulations are nonlinear modulation forms, where the information bits are encoded into the frequency of the modulated signal. During a symbol time, $K = \log_2 M$ bits are encoded into the frequency of the transmitted signal $s(t)$, $0 \le t < Ts$, and $s_i(t) = A\cos(2\pi f_i t + \varphi_i)$ is obtained, where i corresponds to the ith message of log2 M bits and φ_i is the phase associated with the ith carrier frequency.

The simplest case of MFSK is binary FSK (BFSK), where two different carrier frequencies are used for bit 1 and 0 separately. The wave form of BPSK signal is shown in Fig. 4.19.

For MFSK, the modulated signal is given

$$s(t) = \sum_{l=-\infty}^{\infty} g(t - kT_s) A\cos(2\pi f_c t + 2\pi a_i \Delta f_c t + \phi_i) \tag{4.47}$$

where $a_i = (2i - 1 - M), i = 1, 2, \ldots, M = 2^K)$. For a specific symbol time, the modulated signal can be written by

$$s_i(t) = A\cos(2\pi f_c t + 2\pi a_i \Delta f_c t + \phi_i) \quad 0 \le t < T_s \tag{4.48}$$

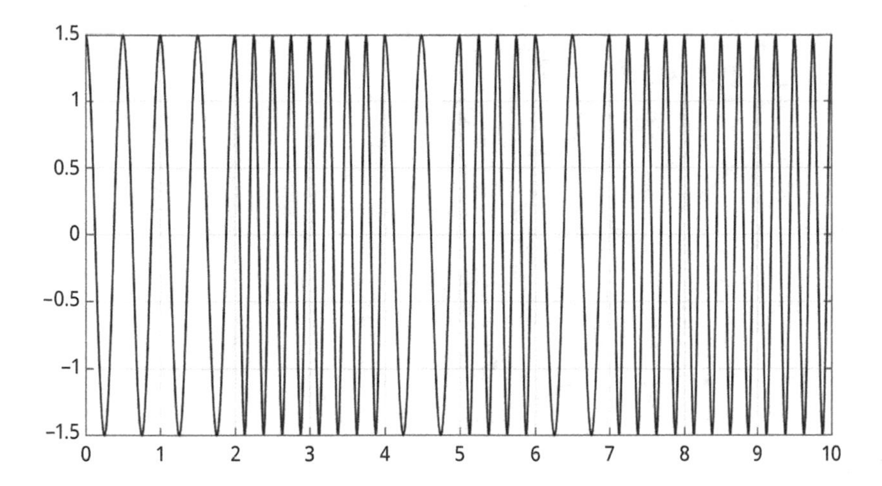

Fig. 4.19: Binary frequency shift keying signal as a function of time.

Apparently, the frequency separation of these carrier frequencies is $2\Delta f$ and the minimum value of $2\Delta f = 1/2T_s$ to guarantee that these basis functions on different frequency are orthogonal. In a given symbol time, only one of the M basis functions is transmitted.

The modulation of MFSK is shown in Fig. 4.20, where M oscillators are operating at the different frequencies $f_i = f_c + \alpha_i \Delta f_c$ and at each symbol time T_s the modulator switches between these different oscillators. Because of the phase offsets between the oscillators, discontinuous phase transition may be caused at the switching times in MFSK implementation, such that an undesirable feature of spectral broadening may be introduced by this phase discontinuity.

Figure 4.21 shows the demodulation of MFSK, where Fig. 4.21(a) is a coherent demodulation structure, and Fig. 4.21(b) is a noncoherent demodulation structure. The coherent demodulation need recover the corresponding phase reference for each of the branches, which is generally difficult and costly. In Fig. 4.21(b), suppose $s_i(t)$ is transmitted, the in-phase branch of the ith arm would be $A_{iI} = A^2\cos^2(\phi_i)$ and the quadrature branch of it would be $A_{iQ} = A^2\sin^2(\phi_i)$. If the noise is ignored, the input to the decision device of the ith branch will be $A^2\cos^2(\Phi_i) + A^2\sin^2(\phi_i) = A^2$, independent of ϕ_i, so that the recovery of the phase is not needed.

4.3.4.2 Minimum shift keying (MSK)

When the frequency separation takes $2\Delta f_c = 1/2T_s$, which is the minimum value that keep the different branches orthogonal, the MFSK is then called minimum shift keying (MSK). The orthogonality of $s_i(t)$ and $s_j(t)$ are guaranteed over a symbol time, for $i \neq j$, and this orthogonality is also required for demodulation. Therefore, this form FSK occupies the minimum bandwidth.

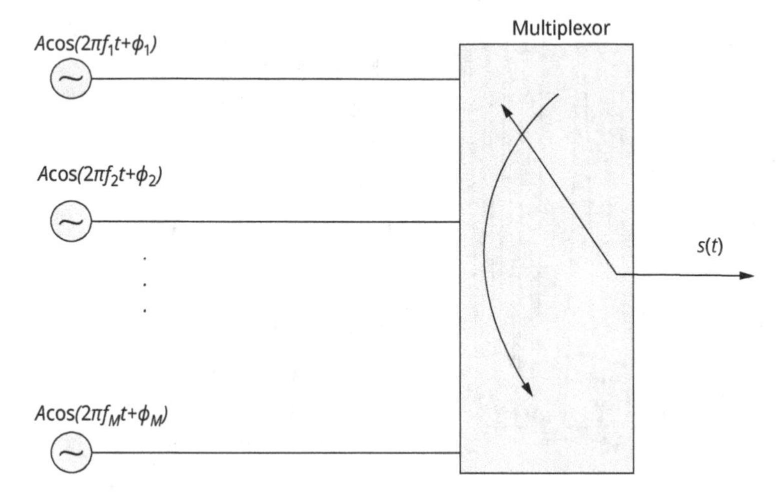

Fig. 4.20: Modulator of MFSK.

4.4 Pulse shaping

As shown in eqs. (4.24), (4.29), (4.44) and (4.47), the bandwidth of the modulated signal is determined by the bandwidth of the pulse shape $g(t)$. Rectangular pulse is the simplest form that makes MPSK a constant envelope. However, this pulse produces very high spectral sidelobes, which means that a larger bandwidth is occupied.

The principle of pulse shaping is that inter-symbol interference (ISI) between pulses in the received signal is not introduced. Let's discuss the main pulse-shaping forms.

4.4.1 Rectangular pulse

Rectangular pulse is the simplest basis pulse,

The expression for $g(t)$ is

$$g(t) = \sqrt{2/T_S}, 0 \le t \le T_S \tag{4.49}$$

In frequency domain

$$G(f, T_S) = \mathcal{F}(g(t)) = \sqrt{2T_s}\ \mathrm{sinc}(\pi f T_s) \exp(-j\pi f T) \tag{4.50}$$

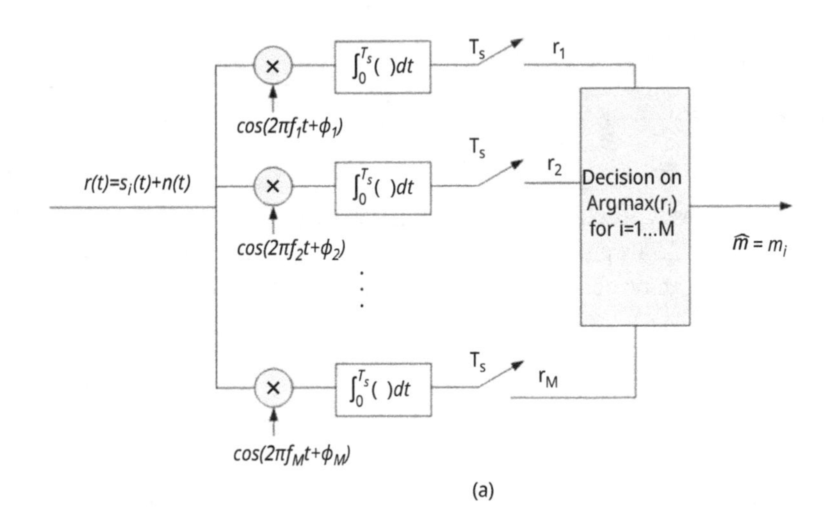

(a)

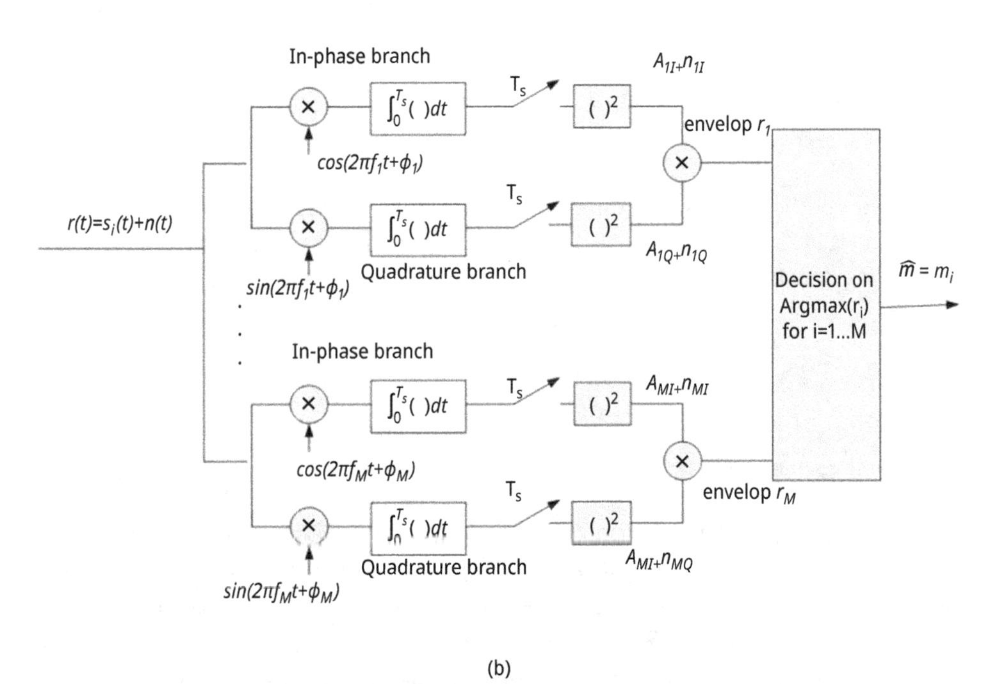

(b)

Fig. 4.21: Demodulation of MFSK. (a) Coherent demodulation of MFSK. (b) Noncoherent demodulation of MFSK.

4.4.2 Cosine pulses

Cosine pulses are written by

$$p(t) = \sin \pi t/T_s \quad 0 \le t \le T_s \tag{4.51}$$

These pulses are mostly used in OQPSK modulation, where the quadrature branch of the PSK modulation has its pulse shifted by $T_s/2$. By using cosine pulses, a constant amplitude modulation can be obtained and the sidelobe energy is 10 dB lower than that of rectangular pulses.

4.4.3 Raised cosine pulse

Raised cosine pulses are designed in the frequency domain according to satisfy a certain spectral requirement. The Fourier Transform of the pulse $p(t)$ is:

$$P(f) = \begin{cases} 1 & |f| < |(1-\beta)/2T \\ \frac{T_S}{2} & \left[1 - \sin\frac{\pi T_S}{\beta}\left(f - \frac{1}{2T_S}\right)\right] \quad (1-\beta)/2T \le |f| \le (1+\beta)/2T \\ 0 & |f| > |(1+\beta)/2T \end{cases} \tag{4.52}$$

where the coefficient β is the rolloff factor determines the rate of spectral rolloff, as shown in Fig. 4.22. And when $\beta = 0$, the spectrum would be a rectangular pulse. The corresponding time domain pulse is

$$p(t) = \frac{\sin \pi t/T_S}{\pi t/T_S} \frac{\cos \beta \pi t/T_S}{1 - 4\beta^2 t^2/T^2{}_S} \tag{4.53}$$

4.4.4 Root cosine pulse

Another form of pulse shaping is called root cosine pulse, which is obtained by taking the square root of the frequency response for the raised cosine. Compared with the raised cosine pulse, the root cosine pulse owns better spectral properties. However, good spectral properties correspond to poor time domain properties. It decays more slowly and this may introduce severe performance degradation because of synchronization errors. It means that it needs more accurate sampling point and a mistiming error in sampling yields a series of intersymbol interference effects. The spectrum is also shown in Fig. 4.22.

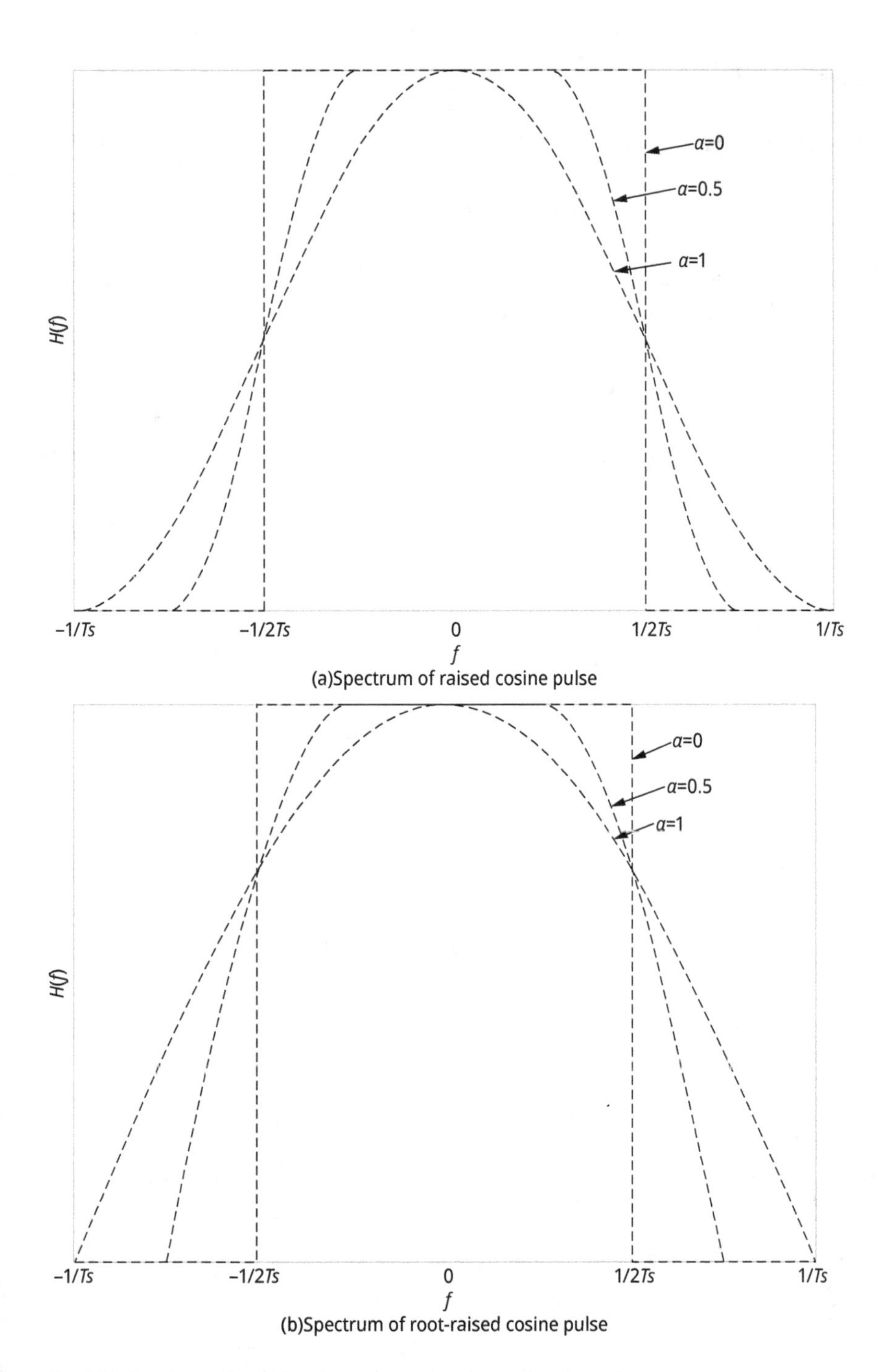

Fig. 4.22: Spectrum of raised cosine pulse and root-raised cosine pulse.

4.4.5 Gaussian pulse

The Gaussian pulse may be used in FSK, and it is defined as

$$g(t) = \frac{\sqrt{\pi}}{\alpha} e^{-\pi^2 t^2/\alpha^2} \tag{4.54}$$

where parameter α denotes the spectral efficiency and is inversely proportional to B_g, the 3 dB bandwidth of $g(t)$, so that it can be written as

$$\alpha = \frac{\sqrt{-\ln\sqrt{0.5}}}{B_g} = \frac{0.5887}{B_g} \tag{4.55}$$

The normalized shape of Gaussian pulses can be shown in Fig. 4.23.

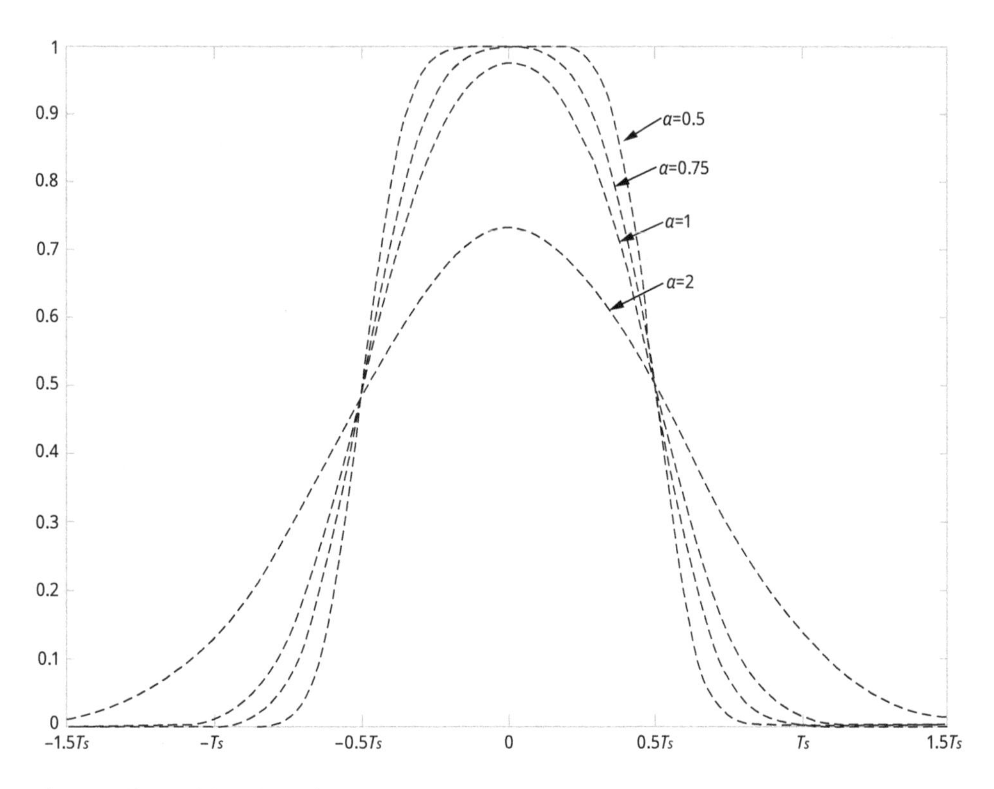

Fig. 4.23: Shape of Gaussian pulses.

The spectrum of $g(t)$ is given by

$$G(f) = e^{-\alpha^2 f^2} \tag{4.56}$$

From Fig. 4.23, one can find that the Gaussian pulse shape does not satisfy the Nyquist criterion, and ISI will be introduced by this pulse shape, which will cause an irreducible error floor from this self-interference. That is why Gaussian pulse applied to MSK (called GMSK) is used for voice in GSM for a relatively high $P_b \approx 10^{-3}$ is tolerable in practice.

4.5 Performance of digital modulations over wireless channels

The relationship between bit error rate and signal-to-noise ratio is discussed, especially in additive Gaussian white noise (AWGN) channel. In the fading channel, signal to noise power ratio (SNR) usually changes randomly with distance or time due to shadow and/or multipath fading.

When the performance of a digital modulation is concerned, we care how the bit error probability varies as a function of SNR, especially in the environment of AWGN. In the fading environment, the SNR is generally random variable over distance or time due to shadowing and/or multipath fading effect, so is the bit error probability. New metrics are needed to describe the performance.

4.5.1 Performance of digital modulations over AWGN channel

4.5.1.1 Signal to noise power ratio and signal to noise power ratio per bit/symbol

In AWGN channel, the received modulated signal can be represented as the addition of the signal and a AWGN noise $n(t)$

$$r(t) = s(t) + n(t) \tag{4.57}$$

Assume $n(t)$ has a power spectral density of $N_0/2$ within the bandwidth of the transmitted signal $s(t)$, the received signal has a power of P_r. Assume the baseband bandwidth is B, the bandwidth of the transmitted signal would be $2B$. Then, the signal to noise power ratio (SNR) can be written as

$$SNR = \frac{P_r}{N_0/2 \times 2B} = \frac{P_r}{N_0 B} \tag{4.58}$$

Let E_b denote the signal energy per bit and E_s denote the signal energy per symbol, the SNR can be rewrite as

$$SNR = \frac{P_r}{N_0 B} = \frac{E_s}{N_0 B T_s} = \frac{E_b}{N_0 B T_b} \tag{4.59}$$

where T_s and T_b are the symbol time and the bit time separately. Let $T_s = k/B$, $SNR = 1/k \times E_s/N_0$, where k is a constant determined by the pulse-shaping filter. For certain pulses with $k = 1$ so that $T_s = 1/B$, for example, the raised cosine pulses with rolloff factor $\beta = 1$, we get $SNR = E_s/N_0$ for multilevel signaling and $SNR = E_b/N_0$ for binary signaling. $\gamma_s = E_s/N_0$ is called SNR per symbol and $\gamma_b = E_b/N_0$ is called SNR per bit.

We typically assume that the symbol energy is divided equally among all bits, then one symbol error corresponds to exactly one bit error when Gray encoding is used and the system is under a reasonable *SNR*s. Assume the size of symbol set is M, then

$$\gamma_s = \gamma_b \log_2 M \tag{4.60}$$

Similarly, we have the relationship between the symbol error probability and bit error probability as,

$$P_s = P_b \log_2 M \tag{4.61}$$

4.5.1.2 Error probability of BPSK

First of all, let's discuss the performance of BPSK with coherent demodulation and perfect carrier frequency and phase synchronization. In BPSK or other binary modulation techniques, the bit error probability exactly equals to the symbol error probability since one bit is also one symbol. According to eqs. (4.31) and (4.32), $s_1(t) = A\cos(2\pi f_c t)$ is for bit 1 and $s_2(t) = -A\cos(2\pi f_c t)$ is for bit 0, where $A > 0$.

According to the definition of Q function, $Q(z)$ equals the probability of a Gaussian random variable x bigger than z where x has a zero mean and unit variance, i.e.

$$Q(z) = p(x > z) = \int_z^\infty \frac{1}{\sqrt{2\pi}} e^{-x^2/2} dx \tag{4.62}$$

Assume the noise is Gaussian distributed with mean 0 and variance $N_0/2$. As shown in Fig. 4.24, suppose bit 0 is sent which corresponds to that the constellation point ($-A$) is sent, an error occurs means the noise is greater than half of the distance between the two constellation points in one direction as lined out in the figure. The case for bit 1 is sent is similar. The bit/symbol error probability of BPSK is

$$P_s = P_b = \frac{1}{2}\sum_{i=0}^{1} p(s_i) = p\left(n > \frac{d_{\min}}{2}\right) = \int_{\frac{d_{\min}}{2}}^{\infty} \frac{1}{\sqrt{\pi N_0}} \exp\left[\frac{-v^2}{N_0}\right] dv = Q\left(\frac{d_{\min}}{\sqrt{2N_0}}\right) \tag{4.63}$$

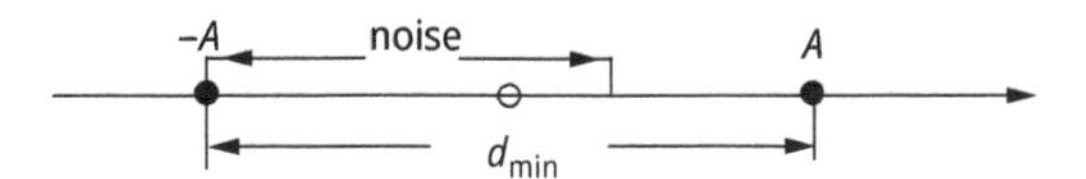

Fig. 4.24: Noise that causes an error in BPSK constellation.

From Fig. 4.24, we obtain $d_{min} = \|A - (-A)\| = 2A$. According to eqs. (4.31) and (4.32), the energy of one bit in BPSK can be represented as a function of A,

$$E_b = \int_0^{T_b} {s_1}^2(t)dt = \int_0^{T_b} {s_2}^2(t)dt = \int_0^{T_b} A^2\cos^2(2\pi f_c t)dt = A^2 \tag{4.64}$$

Similarly, A can be written as a function of E_b,

$$A = \sqrt{E_b} \tag{4.65}$$

And

$$d_{min} = 2A = 2\sqrt{E_b} \tag{4.66}$$

Substitute eq. (4.65) into (4.62), we get

$$P_b = Q\left(\frac{d_{min}}{\sqrt{2N_0}}\right) = Q\left(\frac{2\sqrt{E_b}}{\sqrt{2N_0}}\right) = Q\left(\sqrt{\frac{2E_b}{N_0}}\right) = Q\left(\sqrt{2\gamma_b}\right) \tag{4.67}$$

4.5.1.3 Error probability for coherent demodulations of some typical modulation formats

So far, we have obtained the bit error probability of BPSK. For most modulation formats, it is difficult to obtain the bit error probability or symbol error probability directly. If s_i is transmitted and a symbol error occurs, it means another s_j ($j \neq i$) may be decoded when the noise is greater than half of the distance between s_j and s_i in the direction from s_i to s_j. The symbol s_j ($j \neq i$) which is decoded instead of s_i should be the symbol whose corresponding constellation point has the nearest distance with the received vector. Generally, the upper bound of symbol error probability can be found.

As we have stated in the nearest neighbor approximation eq. (4.21), the symbol error probability P_s can be approximated as the multiplication of probability of error of nearest neighbor constellations at the minimum distance d_{min} and the number of this kind of neighbors M_{dmin}:

$$P_s \approx M_{d_{min}} Q\left(\frac{d_{min}}{\sqrt{2N_0}}\right) \tag{4.68}$$

We can find the relationship between the d_{min} and the energy of a symbol, in such a way, the symbol P_s can be represented as a function of γ_s.

From eq. (4.67), and by using the relationship between P_s and P_b in eq. (4.22) in reasonable SNR and Gray coding as well as the relationship between γ_s and γ_b below

$$\gamma_b = \frac{\gamma_s}{\log_2 M} \tag{4.69}$$

we can obtain the bit error probability P_b as a function of γ_b in coherent demodulations for most of the modulation formats. The detail deduction will not be given, interested readers can refer to the reference (Goldsmith 2005) and the results on approximate symbol/bit error probability for coherent demodulations of some typical modulation formats are cited here:

(1) BPSK

$$P_s = Q\left(\sqrt{2\gamma_s}\right) \tag{4.70a}$$

$$P_b = Q\left(\sqrt{2\gamma_b}\right) \tag{4.70b}$$

(2) QPSK

$$P_s \approx 2Q\left(\sqrt{\gamma_s}\right) \tag{4.71a}$$

$$P_b = Q\left(\sqrt{2\gamma_b}\right) \tag{4.71b}$$

(3) MPSK

$$P_s \approx 2Q\left(\sqrt{2\gamma_s}\sin(\pi/M)\right) \tag{4.72a}$$

$$\mathrm{P}_b \approx \frac{2}{\log_2 M} Q\left(\sqrt{2\gamma_b \log_2 \mathrm{M}}\sin(\pi/M)\right) \tag{4.72b}$$

(4) MPAM

$$P_s \approx \frac{2(M-1)}{M} Q\left(\sqrt{\frac{6\bar{\gamma}_s}{M^2-1}}\right) \tag{4.73a}$$

$$P_b \approx \frac{2(M-1)}{M\log_2 M} Q\left(\sqrt{\frac{6\bar{\gamma}_b \log_2 M}{M^2-1}}\right) \tag{4.73b}$$

(5) 4QAM same as that of QPSK.

(6) Rectangular MQAM

$$P_s \approx \frac{4(\sqrt{M}-1)}{\sqrt{M}} Q\left(\sqrt{\frac{3\bar{\gamma}_s}{M-1}}\right) \tag{4.74a}$$

$$P_b \approx \frac{4(\sqrt{M}-1)}{\sqrt{M}\log_2 M} Q\left(\sqrt{\frac{3\bar{\gamma}_b \log_2 M}{M-1}}\right) \tag{4.74b}$$

(7) Nonrectangular MQAM

$$P_s \approx 4Q\left(\sqrt{\frac{3\bar{\gamma}_s}{M-1}}\right) \tag{4.75a}$$

$$P_b \approx \frac{4(\sqrt{M}-1)}{\sqrt{M}\log_2 M} Q\left(\sqrt{\frac{3\bar{\gamma}_b \log_2 M}{M-1}}\right) \tag{4.75b}$$

Equations (4.70a)–(4.75a) give the general form of symbol error probability as $P_s(\gamma_s) \approx \alpha_M Q(\sqrt{\beta_M \gamma_s})$ and eqs. (4.70b)–(4.75b) give the general form of the bit error probability as $P_b(\gamma_b) \approx \widehat{\alpha}_M Q\left(\sqrt{\widehat{\beta}_M \gamma_b}\right)$, where $\widehat{\alpha}_M = \alpha_M / \log_2 M$ and $\widehat{\beta}_M = (\log_2 M)\beta_M$.

For differential modulation, as stated in the fourth section of 4.3.2, the BER performance is about 3 dB worse than that of the coherent demodulations since the demodulation take the received signal of the current symbol and the previous symbol as input, thus the noise is twice as that of the coherent modulations.

4.5.2 Performance of digital modulations over wireless fading channels

As stated above, the error probability in the AWGA channel is a function of SNR. In fading channels, the received SNR is a random variable due to the variation of power caused by shadowing and/or multipath fading. Suppose the distribution of SNR γ_s is $p_{\gamma s}(\gamma)$, the symbol error probability $P_s(\gamma_s)$ is also random. Three different performance criteria are used to describe the performance of digital modulations over fading channels depend on the rate of change of fading.

1) The outage probability, P_{out}, applies to the slow fading channel. Assume the symbol duration is T_s and the channel coherent time is T_c, when $Ts \ll Tc$, a deep fade will cause large error bursts. In this case, P_{out} is defined as the probability that SNR γ_s falls below a certain threshold, which equivalently corresponds to a maximum allowable P_s.
2) The average error probability, P_s applies to fading channel where the fading is on the same order of a symbol time ($T_s \approx T_c$). In this fairly fast fading case, the fade is roughly at the same value over one symbol time, which means generally one symbol or a few bits can be affected when a fade occurs. This kind of error can be easily corrected by using error correction coding; the bit error performance is generally acceptable. That's why P_s is averaged over the distribution of γ_s for this kind of fading.
3) Outage and average error probability combined metric applies to the case when the channel can be described as fast fading combined with a slow fading, e.g. log-normal shadowing with fast Rayleigh fading.

When the channel is in very fast fading, say $T_c \ll T_s$, the fading will be averaged out by the matched filter in the demodulation and the performance would be the same as in AWGN.

4.5.2.1 Outage probability

The outage probability is defined as the probability the SNR is below a certain target value, say γ_0, which is typical a minimum SNR required for an acceptable performance.

$$P_{out} = p(\gamma_s < \gamma_0) = \int_0^{\gamma_0} p_{\gamma_s}(\gamma) d\gamma \tag{4.76}$$

For example, since human ears are insensitive to digitized voice with bit error rates less than 10^{-3}, SNR corresponding to $P_b = 10^{-3}$ can be set as the target value. From eq. (4.70b), we can find that for a BPSK signal, $P_b = 10^{-3}$ corresponding to $\gamma_0 = 7$ dB. Thus, for a BPSK signal in fading, $\gamma_b < 7$ dB would be declared an outage, so we set $\gamma_0 = 7$ dB.

In Rayleigh fading channel, substitute the probability density function (pdf) of Rayleigh distribution into eq. (4.76), we get the outage probability becomes

$$P_{out} = \int_0^{\gamma_0} \frac{1}{\overline{\gamma_s}} e^{-\frac{\gamma_s}{\overline{\gamma_s}}} d\gamma_s \tag{4.77}$$

From eq. (4.77), the required average SNR $\overline{\gamma_s}$ can also be deduced if the outage probability is given.

$$\overline{\gamma_s} = \frac{\gamma_0}{-\ln(1 - P_{out})} \tag{4.78}$$

4.5.2.2 Average probability of error

The average probability of error applies in the situation $T_s \approx T_c$. In this case, we assume the SNR per symbol γ_s is roughly unchanged during one symbol time so that the average error probability can be obtained by integrating the error probability in AWGN over the fading distribution

$$\overline{P}_s = \int_0^{\infty} P_s(\gamma) p_{\gamma_s}(\gamma) d\gamma \tag{4.79}$$

where $P_s(\gamma)$ is the symbol error probability as a function of γ in AWGA channel, and the approximations for typical modulation formats can be found from eqs. (4.70a) to (4.75a). $p_{\gamma_s}(\gamma)$ is the distribution of the received SNR.

For example, for a BPSK modulation, suppose in 50% of time the signal to noise power ratio per bit γ_b is 13 dB, and in the rest of time γ_b is $-\infty$, then we can find the average signal to noise power ratio per bit $\overline{\gamma}_b$ is 10 dB. The bit error probability corresponding to the two states is 10^{-9} and 0.5, and the average bit error probability is $\overline{P}_b = \sum_{i=1}^{2} P_b(\gamma_i) p(\gamma_i) = 10^{-9} \times 0.5 + 0.5 \times 0.5 = 0.25$. As we know, the bit error probability of BPSK is 2×10^{-5} in AWGN channel when γ_b is 10 dB.

For the general case, the first step is to obtain the distribution $p_{\gamma_s}(\gamma)$ from the distribution of amplitude. Suppose the amplitude of the received signal is Rayleigh distributed,

$$p(r) = \frac{r}{a^2} e^{-\frac{r^2}{2\sigma^2}}, r \geq 0 \tag{4.80}$$

The instantaneous received power $P_{inst} = r^2$and the average power $\bar{P} = 2\sigma^2$. The pdf of the received power is derived by using the Jacobian $|dP_{inst}/dr = 2r|$,

$$p(P_{inst}) = \frac{1}{\bar{P}} e^{-\frac{P_{inst}}{\bar{P}}}, P_{inst} \geq 0 \tag{4.81}$$

Since the SNR is the received power divided by the noise power, the pdf of the SNR is thus

$$p(\gamma_s) = \frac{1}{\bar{\gamma}_s} e^{-\frac{\gamma_s}{\bar{\gamma}_s}} \tag{4.82}$$

where $\bar{\gamma}$ is the average SNR per symbol.

Substitute eq. (4.81) into (4.79), by using

$$2\int_0^\infty Q(\sqrt{2x}) a e^{-ax} dx = 1 - \sqrt{\frac{1}{1+a}} \tag{4.83}$$

the average error probability of bit for coherent binary phase shift keying (BPSK) can be obtained,

$$\bar{P}_b = \frac{1}{2}\left[1 - \sqrt{\frac{\bar{\gamma}_b}{1+\bar{\gamma}_b}}\right] \approx \frac{1}{4\bar{\gamma}_b} \quad \text{(BPSK)} \tag{4.84}$$

For coherent demodulation of BFSK, we can find the average probability of bit error through a similar way,

$$\bar{P}_b = \frac{1}{2}\left[1 - \sqrt{\frac{\bar{\gamma}_b}{2+\bar{\gamma}_b}}\right] \approx \frac{1}{2\bar{\gamma}_b} \quad \text{(BFSK)} \tag{4.85}$$

The average probability of bit error of DPSK is

$$\bar{P}_b = \frac{1}{2(1+\bar{\gamma}_b)} \approx \frac{1}{2\bar{\gamma}_b} \quad \text{(DPSK)} \tag{4.86}$$

The above approximations are based on reasonably large SNR assumption. It can be found that comparing the coherent demodulation of BPSK, the performance of DPSK has a 3 dB loss in Rayleigh fading channel, which is similar to the case in the AWGN channel.

For the general expression $P_s(\gamma_s) \approx \alpha_M Q(\sqrt{\beta_M \gamma_s})$, the approximation of the symbol error probability in Rayleigh fading is

$$\bar{P}_s \approx \int_0^{\infty} \alpha_M Q\left(\sqrt{\beta_M \gamma_s}\right) \frac{1}{\bar{\gamma}_s} e^{-\frac{\gamma_s}{\bar{\gamma}_s}} d\gamma_s = \frac{\alpha_M}{2}\left[1 - \sqrt{\frac{0.5\beta_M \bar{\gamma}_s}{1 + 0.5\beta_M \bar{\gamma}_s}}\right] \approx \frac{\alpha_M}{2\beta_M \bar{\gamma}} \tag{4.87}$$

The approximation for bit error probability in reasonable SNR can be obtained by eq. (4.68).

Of course, if the channel is not Rayleigh, it would be more complicated. From eqs. (4.83) to (4.86) and eqs. (4.70) to (4.75), we can find out that there is a big gap in the performance of digital modulations under AWGN and Rayleigh fading channel. The BER of BPSK and MQAM decreases exponentially with the increase of SNR per bit γ_b in the AWGN channel, while the average BER of BPSK and MQAM decreases linearly with the increase of average SNR per $\bar{\gamma}_b$ in the Rayleigh fading channel. For a given BER, especially a very small BER, the average power required for the Rayleigh fading channel is much higher than that for the AWGN channel. For BPSK in the AWGN channel, a γ_b of about 7 dB is required to maintain a BER of 10^{-3}, while an average SNR per $\bar{\gamma}_b$of 24 dB is required to maintain an average BER of 10^{-3} in Rayleigh fading channel. Fortunately, several techniques, such as diversity and spread spectrum with RAKE receiver, have been developed to compensate the fading effects, and these techniques will be discussed in the following chapters. In the Ricean fading channel, it is difficult to obtain closed form solution, interested users may check the reference (Bello et al. 1962) for the bounds on the bit-error probabilities of 2DPSK and 4DPSK in the Ricean fading situation.

4.5.2.3 Combined outage probability and average probability of error

The outage probability applies to a slow fading channel, and the average error probability applies to the fading that is approximately at the order at the symbol time. When fast fading superimposes over slow fading, we use the metric of combined outage probability and the average probability of error to represent the performance. The outage probability is used to give the probability that the slow fading (averaged over fast fading) falls below a certain threshold. In non-outage state ($\bar{\gamma}_s \geq \bar{\gamma}_{s0}$), the average probability of error is calculated over fast fading, i.e.

$$\bar{P}_s = \int_0^{\infty} P_s(\gamma_s) p_{\gamma_s}(\gamma_s | \bar{\gamma}_s) d\gamma_s \tag{4.88}$$

where $\bar{\gamma}_s$ is average SNR per symbol, but this averaging is only over fast fading, it is still a random variable depends on fixed path loss and random shadowing.

A target average SNR per symbol $\bar{\gamma}_{s0}$ is determined by an acceptable performance threshold which is measured by a maximum average probability of symbol error, where $\overline{P}_{s0}$ satisfies

$$\overline{P}_{s0} = \int_0^\infty P_s(\gamma_s) p_{\gamma_s}(\gamma_s | \bar{\gamma}_{s0}) d\gamma_s \tag{4.89}$$

4.5.3 Effect of delay and frequency spread on the error probability

4.5.3.1 Irreducible errors floors and their physical causes

In AWGN channels, transmission errors are caused mainly by noise, in which the bit error probability is generally decreased exponentially as the increase of SNR. However, in wireless propagation channels, in addition to noise, there are signal distortions that may cause errors, in which the bit error probability at a certain level will not be decreased with the increase of transmit power. This irreducible error level is often called irreducible error floor.

These distortions are created by delay spread (i.e., multipath components arriving with different delays) or frequency spread (i.e., signal components arriving with different Doppler frequency shifts). The delay spread is the key factor affecting the performance in high data rates applications, whereas the Doppler spread is the key factor in low data rates applications.

The so-called irreducible errors can also be eliminated by other methods such as equalization, diversity and so on instead of increasing power. These techniques to reduce the error will be treated in later chapters, and we only discuss how the spread in delay and/or frequency affects the performance of digital modulations.

4.5.3.2 Doppler spread

Apparently, when the information is encoded in the frequency and related to the frequency, the Doppler spread or frequency dispersion may cause errors. For example, it is obvious that a big enough frequency shift of the received signal of FSK can push a bit over the decision boundary and this will cause errors.

Another scenario is the differential modulations. As we have discussed, differential detection assumes that the channel remains unchanged between two adjacent symbols, so that the phase of the previous symbol acts as a reference, and the phase difference of the two adjacent symbols is used to convey information. Several researches have been done (Bello et al. 1962, Kam 1998, Schwartz et al. 1966, Proakis 2000, Simon et al. 2000) to find the relationship between the bit error probability and the Doppler shift. As a result, the bit error probability for DPSK can be written as (Goldsmith 2005),

$$\bar{P}_b = 0.5\left[\frac{1+K+\bar{\gamma}_b(1+\rho_C)}{1+K+\bar{\gamma}_b}\right]\exp\left(-\frac{K+\bar{\gamma}_b}{1+K+\bar{\gamma}_b}\right) \tag{4.90}$$

where ρ_C is the channel correlation coefficient at distance of a bit time T_b, K is the fading parameter of the Ricean distribution, $\bar{\gamma}_b$ is the average SNR per bit. When $K = 0$, the channel is reduced to Rayleigh fading channel, and the average bit error probability is simplified to

$$\bar{P}_b = 0.5\left[\frac{1+\bar{\gamma}_b(1+\rho_C)}{1+\bar{\gamma}_b}\right] \tag{4.91}$$

For Ricean and Rayleigh fading channels, let $\bar{\gamma}_b \to \infty$ in (5.90) and (5.91) separately, we get the irreducible error floor for Ricean and Rayleigh fading channels

DPSK Ricean fading channel:

$$\bar{P}_{floor} = 0.5(1-\rho_C)e^{-K} \tag{4.92}$$

DPSK Rayleigh fading channel:

$$\bar{P}_{floor} = 0.5(1-\rho_C) \tag{4.93}$$

By a similar way, we can obtain the bound of DPSK in Ricean fading channel (Kam 1998), by letting $\bar{\gamma}_b \to \infty$, the irreducible symbol error floor of DQPSK is

DQPSK Ricean fading channel:

$$\bar{P}_{floor} = 0.5\left[1-\sqrt{\frac{(\rho_C/\sqrt{2})^2}{1-(\rho_C/\sqrt{2})^2}}\right]\exp\left[-\frac{(2-\sqrt{2})(K/2)}{1-\rho_C/\sqrt{2}}\right] \tag{4.94}$$

DQPSK Rayleigh fading channel:

$$\bar{P}_{floor} = 0.5\left[1-\sqrt{\frac{(\rho_C/\sqrt{2})^2}{1-(\rho_C/\sqrt{2})^2}}\right] \tag{4.95}$$

Assuming the uniform scattering model ($\rho_C = J_0(2\pi f D T_b)$) in eq. (4.93), we obtain the irreducible error for DPSK Rayleigh fading channel.

DPSK Rayleigh fading channel:

$$\bar{P}_{floor} = 0.5(1-J_0(2\pi f_D T_b)) \approx 0.5(\pi f_D T_b)^2 \tag{4.96}$$

In the above equation, we can find that the error floor is a function of the product of f_D and T_b. For a given communication system, if the Doppler shift is determined, a small bit time, or equivalently, a higher bit rate can guarantee a lower error floor.

4.5.3.3 Delay spread and inter symbol interference

Compared with the Doppler spread, the delay dispersion has a great impact on the high-data-rate wireless communications. Inter-symbol interference (ISI) is affected by the ratio of delay spread to the symbol duration. The maximum excess delay of a channel impulse response is a parameter of the channel and depends on the operation environment, not the communication system.

Let's assume that the maximum excess delay is 1 μs and the symbol duration of a communication system is 10 μs, then the ISI can disturb 10% of a symbol. If the symbol duration of another system is 5 μs, it will disturb 20% of the symbol.

Many analyses and experimental results (Chuang 1987, Gurunathan et al. 1992) suggested that the error floor caused by the ISI can be written as

$$\bar{P}_{floor} = K(\sigma_{T_M}/T_s)^2 \tag{4.97}$$

where σ_{T_M} is the rms delay spread and T_s is the symbol time, the constant K is dependent on the modulation format, the transmitter and receiver filters, the form of the average impulse response and choice of the sampling instant. Since σ_{T_M} is determined by the channel, if we want the probability of error small enough so that the performance is acceptable, the lower bound of T_s as well as the upper bound of the symbol rate is determined. If the order of modulation is given, the date rate is limited.

Problems

4.1 Figure out the advantages of digital modulations over analogy modulations.

4.2 In designing a communication system, what are the main considerations in selecting a modulation format and demodulation scheme?

4.3 For a bit sequence 101011001, assume Gray coding is used and the starting symbol is the kth symbol with the transmitted symbol at the $(k-1)$th symbol time $s(k-1) = A\exp(j\pi/2)$, find the symbol sequence transmitted by using differential QPSK modulation with Gray encoding.

4.4 What is the minimum frequency separation for FSK and show why?

4.5 Assume the average signal power of a communication system based on BPSK modulation is −40dBm and the noise power within the signal bandwidth is −60dBm at the receiver. Is this system suitable for the transmission of data? How about voice? How about if we change the channel into a Rayleigh fading channel with an average SNR $\overline{\gamma_b} = 20$ dB?

4.6 Consider a BPSK modulation system working in rural environments with a channel delay spread $\sigma_{Tm} = 25$ μs, and find the maximum data rate that can be supported if the probability of bit error $P_b < 10^{-3}$ is required. The error floor $\bar{P}_{floor} \le (\sigma_{T_M}/T_s)^2$ can be used for calculation. How about QPSK and 8PSK? If the modulation systems work in the urban environment where $\sigma_{Tm} \approx 2.5$ μs, how about the above question?

Chapter 5
Diversity, channel coding and equalization

From Chapters 2 to 4, we know that there are various types of performance degradation due to wireless channels and the wireless link might be deteriorated seriously. These signal propagation effects, such as path loss, shadowing and multipath, especially Doppler spread and delay spread have very strong impacts on the bit error rate of the modulation techniques. The received signal of mobile wireless communications may be distorted or faded significantly compared to the received signal under AWGN channels.

Diversity, channel coding and equalization are three techniques which can be used to improve the link quality of wireless channels.

Diversity is an effective fading channel compensating technique, which is implemented by two or more independent paths. The independent paths may be implemented in time, space, frequency and polarization domains. As the paths are statistically independent, the probability that all paths are in a deep fading simultaneously is low. Combining these independent paths, a higher-received SNR is obtained.

Channel coding is another technique to improve the wireless link performance by adding error detection or error correction bits in the transmitted message. The code adding error detection bits are called error detection codes, whereas those adding error correction bits are called error correction codes. Since decoding is a processing after demodulation, coding is regarded as a postdetection technique. Of course, the added redundant bits will generally lower the raw data transmission rate or expand the occupied bandwidth with a particular data rate, although the trellis coded modulation which combines the coding and modulation achieves coding gain without bandwidth expansion, it is not popular.

Equalization is a technique to compensate the intersymbol interference (ISI) introduced by the multipath effects implemented at the receiver. As stated in Chapters 3 and 4, when the signal has a bandwidth greater than the channel coherence bandwidth, the signal experiences frequency selective fading and the modulated symbol spreads in time. Equalizer is then used to mitigate the impact of delay spread. Of course, techniques used at transmitter to make the signal less susceptible to delay spread can also be used such as spread spectrum and multicarrier modulation, which will be discussed in the later chapters.

https://doi.org/10.1515/9783110751437-005

5.1 Diversity

5.1.1 Different realization methods of diversity

The fundamental of diversity is to realize independent fading paths. Independence means that the joint probability density function (pdf) of field strength (or power) of these paths equals the product of marginal pdfs for the channels, that is

$$\mathrm{pdf}_{r_1,r_2,\ldots}(r_1,r_2,\ldots)=\mathrm{pdf}_{r_1}(r_1)\mathrm{pdf}_{r_1}(r_1)\ldots \tag{5.1}$$

If the *correlation coefficient* that characterizes the correlation between these paths is not zero, the effectiveness of diversity will be decreased.

The definition of correlation coefficient of signal envelopes x and y is

$$\rho_{xy}=\frac{E\{x\cdot y\}-E\{x\}\cdot E\{y\}}{\sqrt{\left(E\{x^2\}-E\{x\}^2\right)\cdot\left(E\{y^2\}-E\{y\}^2\right)}} \tag{5.2}$$

The correlation coefficient would be zero when x and y are independent since $E\{xy\} = E\{x\}E\{y\}$ in the independent case. However, when ρ is below a certain value, say 0.5 or 0.7, we regard that the signals are "effectively" decorrelated.

To realize different independent or "effectively" decorrelated paths, several diversity strategies are used.

Firstly, by using multiple transmit or receive antennas, also called antenna array, independent paths can be implemented in space domain. With receiver space diversity, independent paths are realized without increasing the signal power and bandwidth, but the transmit power would be divided among multiple antennas in transmitter diversity. Moreover, with coherent combining of receiver space diversity, the receive SNR will be increased over the SNR that would be obtained with just a single antenna at receiver. For the transmitter space diversity, the received SNR is the same as if there were just a single transmit antenna. It should be noted that space diversity requires the separation between antennas so that the fading corresponding to each antenna is nearly independent. Theoretically, the minimum antenna separation on each antenna for independent fading is approximately one half (0.38λ to be exact) wavelength in the uniform scattering environment (Goldsmith 2005). If the transmit or receive antennas are directional, a larger antenna separation is required to get independent fading paths.

The second kind of independent paths can be implemented by using two antennas with two different polarizations, that is, a vertically polarized wave and a horizontally one. Because the scattering angle relative to each polarization is random, the probability that the two signals received on the differently polarized antennas are in deep fades simultaneously would be very low. The disadvantage is that the power is divided by the two paths in the transmission.

Thirdly, independent paths can also be implemented by using directional antennas where the received antenna beam width is restricted to a given angle. For a very narrow angle, it is possible to allow only one ray falls within that receive beam width, and there will be no multipath fading. There are two kinds of implementations. One is to use a large number of directional antennas to span all directions of arrival and the other is to use a single antenna to steer the directivity to the angle of arrival so that only one multipath component (preferably the strongest one) is received. Smart antennas are regarded as this kind of techniques, where the phases of the antenna elements are adjusted so that the directional antennas are formed to be steered to the incoming angle of the strongest multipath component.

Fourthly, independent paths can also be implemented in frequency domain such as multicarrier modulations, where the carriers are separated by the channel coherence bandwidth. The typical implementing method is the OFDM. Spread spectrum with RAKE receiver can also be regarded as a kind of frequency diversity.

Finally, independent paths can also be implemented in time domain by transmitting the same bit sequence in two or more time slots separated by the channel coherence time. Coding combined with interleaving is one kind of time diversity.

In this chapter, we take space diversity as an example to describe the principle of diversity and the combining techniques. Of course, the combining techniques can also be applied to any type of diversity.

5.1.2 Combining techniques of receiver diversity

5.1.2.1 Array gain and diversity gain

Generally, the combiner is a weighted sum of the different fading branches. The combining of the signals of more than one branches requires cophasing. Otherwise, the signals would add up uncoherently and the yielding output could still exhibit significant fading effect because the addition of the signals in different branches would be constructive and destructive. Even in the absence of fading, the combining can lead to an increase in average-received SNR. Suppose the average SNR of one branch is $\bar{\gamma}$, the average SNR of the combiner is $\bar{\gamma}_\Sigma$, then the array gain A_g is defined as

$$A_g = \frac{\bar{\gamma}_\Sigma}{\bar{\gamma}} \tag{5.3}$$

The outage probability P_{out} defined by eq. (4.76) and average error probability $\bar{P}_s$ defined by eq. (4.79) can be then found out by the combined SNR γ_Σ. The $\bar{P}_s$ and P_{out} are decreased because of the diversity combination, and the resulting performance advantage is called the diversity gain. Let's take the typical average probability of error representation $\bar{P}_s = c\bar{\gamma}^{-M}$, where the constant c depends on the specific modulation format and coding, the average received SNR $\bar{\gamma}$ is averaged over all branches and M is the diversity order. As we have discussed in Chapter 4 that the average error probability in Rayleigh fading with no diversity can be expressed proportional to $\overline{\gamma}^{-1}$, which has a diversity order of one. The diversity orders a system can achieve at most with M antennas is M, and the system is said to achieve full diversity orders when the order is M.

Now let's discuss the combination of signals from different antennas. For simplification, we mainly discuss combining signals from multiple antennas mounted at receiver. Of course, the techniques are valid for other types of diversity signals mathematically. In general, there are three typical combining methods, selection combining (SC), maximal ratio combining and equal gain combining.

5.1.2.2 Selection combining

SC technique selects the branch with the highest SNR as the result diversity output and discards other branches. In practice, it is not convenient to detect the SNR of all the branches all the time, the highest instantaneous signal power (or received signal strength indication – RSSI) is used instead. Suppose there are M antenna elements, M RSSI sensors, and M-to-1 multiplexer is used. We will track the RSSI of all the branches and switch to the new best antenna when its RSSI becomes the highest. Figure 5.1 shows the implementation of the SC.

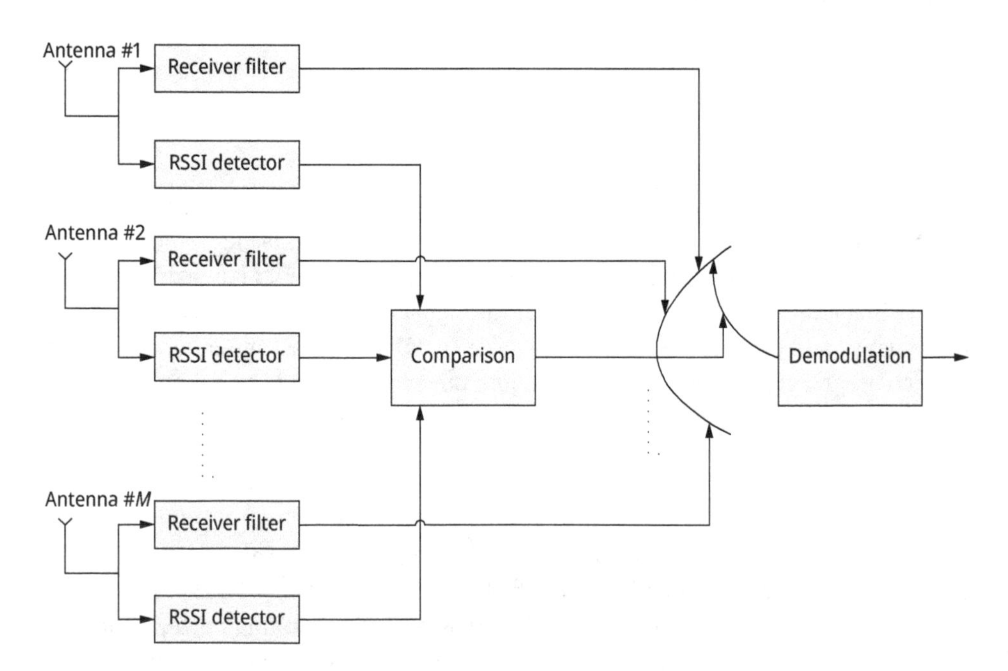

Fig. 5.1: Implementation of selection combining.

Since only one path with the maximum SNR is output, cophasing is not required. The CDF of γ_Σ for M branch selection combiner is

$$P_{\gamma_\Sigma}(\gamma) = p(\gamma_\Sigma < \gamma) = p\left(\max_{i=1}^{M}[\gamma_i < \gamma]\right) = \prod_{i=1}^{M} p(\gamma_i < \gamma) \tag{5.4}$$

For Rayleigh fading amplitudes channel, the SNR distribution is exponential:

$$p(\gamma_i) = \frac{1}{\overline{\gamma}_i}\exp\left(-\frac{\gamma_i}{\overline{\gamma}_i}\right) \tag{5.5}$$

Then the cumulative distribution function (cdf) is

$$\mathrm{cdf}_{\gamma_i}(\gamma_i) = 1 - \exp\left(-\frac{\gamma_i}{\overline{\gamma}_i}\right) \tag{5.6}$$

From eq. (4.77), we know that the outage probability for a target γ_0 is

$$P_{\mathrm{out}}(\gamma_0) = 1 - \exp\left(-\frac{\gamma_0}{\overline{\gamma}_i}\right) \tag{5.7}$$

Assume all the M branches are identical, and the average branch SNR is $\overline{\gamma}$. We can find the cdf of the selection combiner as

$$\mathrm{cdf}_{\gamma_\Sigma}(\gamma_\Sigma) = \left[1 - \exp\left(-\frac{\gamma_i}{\overline{\gamma}}\right)\right]^M \tag{5.8}$$

By definition, the outage probability of the selection-combiner for identical branches and target γ_0 is then

$$P_{\mathrm{out}}(\gamma_0) = \left[1 - \exp\left(-\frac{\gamma_0}{\overline{\gamma}}\right)\right]^M \tag{5.9}$$

5.1.2.3 Threshold combining and switch and stay combining

The system with SC requires a dedicated receiver on each branch to continuously monitor RSSI, make a comparison and control the switching when the best branch changes.

Threshold combining (TC) is a simpler version of SC. The system with TC scans each of the branches in sequential order and output the first signal with RSSI above a given threshold. When the RSSI of the currently active branch falls below the threshold, the system will restart the scanning in sequential order and output the new found first signal with RSSI above the threshold.

If there are only two-branches exist, the principle of the TC becomes switching to the other branch when the RSSI on the active branch falls below a given threshold. In this case, the method becomes the so-called switch and stay combining (SSC). A key problem of SSC is that the RSSI of both branches may all fall below the threshold. We need carefully consider the choosing of threshold in TC and SSC so that it's not too low or too high. If it is too low, the active branch is kept even when the other branch might offer better quality; if it is too high, the branch might have to switch to other branch even the branch to be switched to actually offer lower signal quality than the currently active one.

5.1.2.4 Maximum ratio combining

In SC and SSC, only one branch of the signals is output to the demodulator. The output of maximum ratio combining (MRC) diversity is a weighted sum of all branches. Since the signals are cophased, the phases corresponding to the incoming signals on each of the branches are compensated. Thus, we assume the input signals from the M branches are $r_1 e^{j\theta_1} s(t), r_2 e^{j\theta_2} s(t), \ldots, r_M e^{j\theta_M} s(t)$, the weights for the M branches are $\alpha_1 = a_1 e^{-j\theta_1}, \alpha_2 = a_2 e^{-j\theta_2}, \ldots, \alpha_M = a_M e^{-j\theta_M}$, where the $e^{-j\theta_i}, i = 1, 2, \ldots, M$ is used to compensate the phases. Then, the envelop of the cophased combiner is

$$r = \sum_{i=1}^{M} a_i r_i \tag{5.10}$$

Assume the PSD of noise on all branches are the same N_0, the PSD of the noise at the diversity combiner is

$$N_{\text{total}} = \sum_{i=1}^{M} a_i^2 N_0 \tag{5.11}$$

Then, the SNR at the diversity output is

$$\gamma_\Sigma = \frac{r^2}{N_{\text{total}}} = \frac{\left(\sum_{i=1}^{M} a_i r_i\right)^2}{\sum_{i=1}^{M} a_i^2 N_0} \tag{5.12}$$

According to the Cauchy-Schwartz inequality, the maximum value for eq. (5.12) occurs at $a_i^2 = r_i^2 / N_0$. Substitute the optimal weights into eq. (5.12), the optimal combiner output SNR is

$$\gamma_\Sigma = \frac{r^2}{N_{\text{total}}} = \frac{\left(\sum_{i=1}^{M} r_i \cdot r_i\right)^2}{\sum_{i=1}^{M} r_i^2 N_0} = \sum_{i=1}^{M} r_i^2 / N_0 = \sum_{i=1}^{M} \gamma_i \tag{5.13}$$

The principle of MRC is shown in Fig. 5.2.

For statistical channel models, if channels are independent to each other, we can find the moment-generating function of the total SNR by computing the product of the characteristic functions of the branch SNRs. If each of the branches is Rayleigh distributed, the distribution of SNR would be exponential. Given each of the branches has a

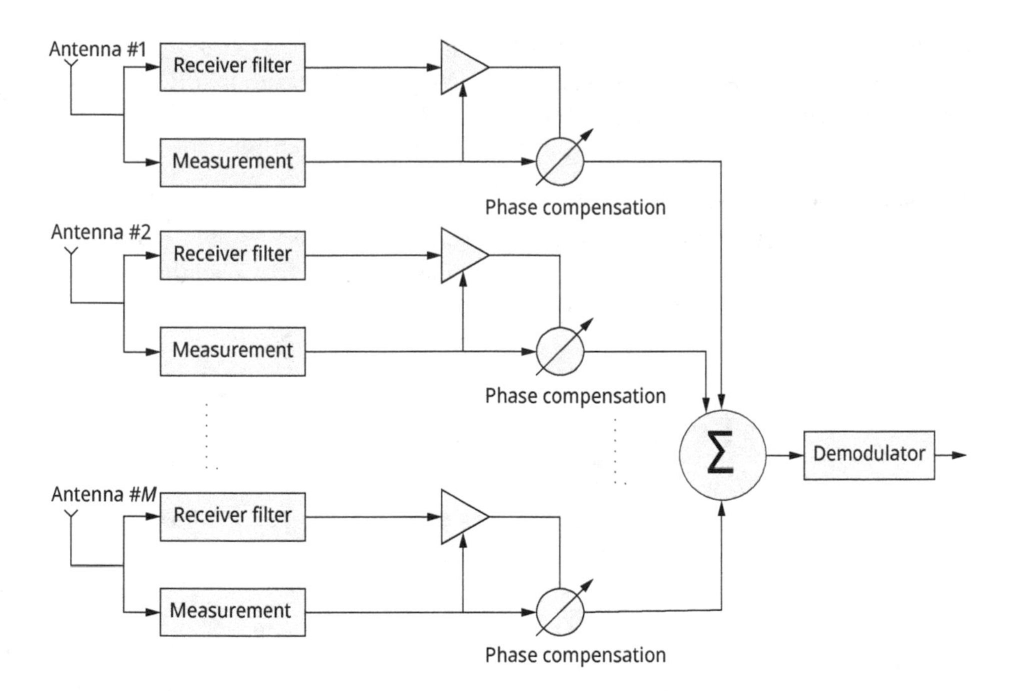

Fig. 5.2: Principle of maximum ratio combining.

same mean SNR of $\overline{\gamma_i} = \overline{\gamma}$, $i = 1, 2, \ldots, M$, after several manipulation operations, we can obtain the pdf of the MRC output:

$$p_{\gamma_\Sigma}(\gamma) = \frac{\gamma^{M-1} e^{-\gamma/\overline{\gamma}}}{\overline{\gamma}^M (M-1)}, \gamma \geq 0 \tag{5.14}$$

The mean SNR of MRC output is

$$\overline{\gamma_\Sigma} = M\,\overline{\gamma} \tag{5.15}$$

The outage probability for a given target value γ_0 is

$$P_{\text{out}} = p(\gamma_\Sigma < \gamma_0) = \int_0^{\gamma_0} p_{\gamma_\Sigma}(\gamma) d\gamma = 1 - e^{-\gamma_0/\overline{\gamma}} \sum_{k=1}^{M} \frac{(\gamma_0/\overline{\gamma})^{k-1}}{(k-1)!} \tag{5.16}$$

It is difficult to obtain the general expression for average error of symbol, for high SNR and the identically distributed γ_i, $i = 1, 2, \ldots M$ with $\overline{\gamma_i} = \overline{\gamma}, i = 1, 2, \ldots, M$, the average error of symbol is

$$\overline{P_s} \approx \alpha_M \left(\frac{\beta_M \overline{\gamma}}{2} \right)^{-M} \tag{5.17}$$

where α_M and β_M are same coefficients as those in eqs. (4.70a)–(4.75a). As we discussed before, M is the number of antennas, and eq. (5.17) indicates that full diversity order is achieved in the MRC technique.

5.1.2.5 Equal gain ratio combining

Equal gain ratio combining (EGC) is simpler than MRC because it is not required to know the time-varying SNR on each branch. Assume the noise PSD on all the branches are at the same level, the cophased signals on each branch are combined with equal weighting:

$$\gamma_\Sigma = \frac{\left(\sum_{i=1}^{M} r_i\right)^2}{N_0 M} = \frac{\left(\sum_{i=1}^{M} r_i/\sqrt{N_0}\right)^2}{M} = \frac{\left(\sum_{i=1}^{M} \sqrt{\gamma_i}\right)^2}{M} \tag{5.18}$$

If all branches suffer from Rayleigh fading with the same mean SNR $\overline{\gamma}$, the mean output SNR of the EEG diversity can be found in Molisch (2011)

$$\overline{\gamma_\Sigma} = \overline{\gamma}\left(1 + (M-1)\frac{\pi}{4}\right) \tag{5.19}$$

Compared to eq. (5.15), we can draw a conclusion that the performance of EGC is worse than that of MRC by only a factor $\pi/4$ (in terms of mean SNR). If the mean SNRs of the branches are also different, the performance gap between the two combining techniques becomes bigger.

The principle of the EEG is shown in Fig. 5.3.

The pdf and CDF of γ_Σ for EEG generally do not have a closed-form solution. If there are two i.i.d. Rayleigh fading branches, closed-formed solutions do exist. The outage probability and average probability of bit error for BPSK are given in eqs. (5.20) and (5.21) (Yacoub 1993):

$$P_{\text{out}}(\gamma_0) = 1 - e^{-2\gamma_0/\overline{\gamma}} - \sqrt{\pi\gamma_0/\overline{\gamma}}e^{-\gamma_0/\overline{\gamma}}\left(1 - 2Q\left(\sqrt{2\gamma_0/\overline{\gamma}}\right)\right) \tag{5.20}$$

$$\overline{P_b} = \int_0^\infty Q\left(\sqrt{2\gamma}\right)p_{\gamma_\Sigma}(\gamma)d\gamma = 0.5\left(1 - \sqrt{1 - \left(\frac{1}{1+\overline{\gamma}}\right)^2}\right) \tag{5.21}$$

It should be mentioned that the conclusion we drew that MRC performs the best among the combining techniques is based on the assumption that only AWGN disturbs the signal. If there is interference that dominates the signal quality, the MRC is no longer the optimal solution. The weights should be determined in order to maximize the signal-to-interference-and-noise ratio. This strategy is called optimum combining, and interested readers can refer reference (Molisch 2011, Winters 1984).

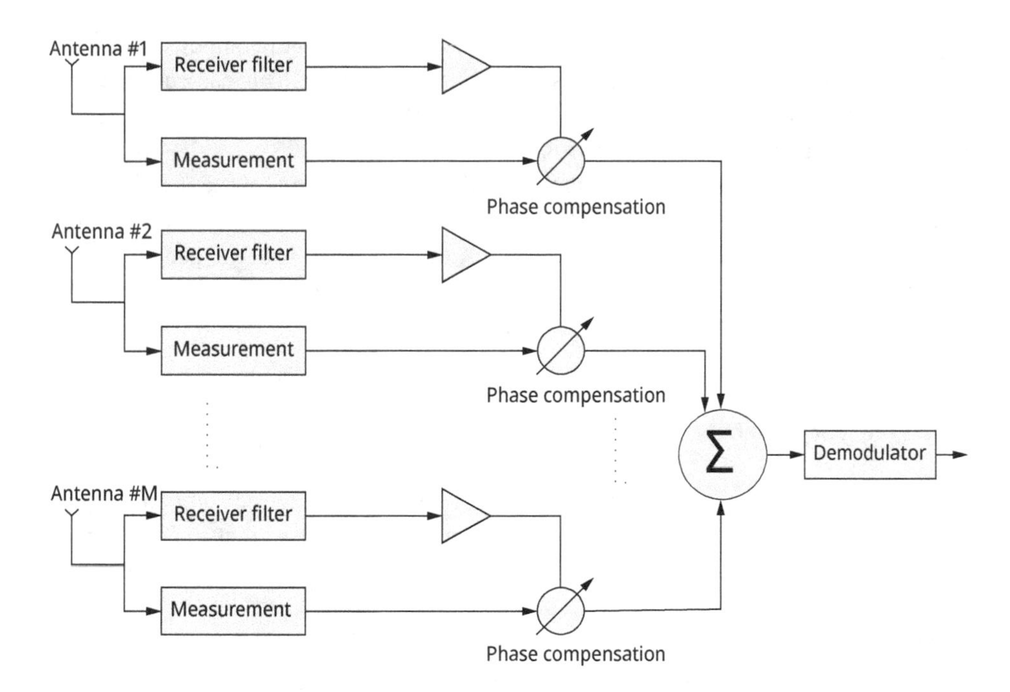

Fig. 5.3: Principle of equal gain combining.

5.1.3 Polarization diversity

Space diversity is less practical in the case of narrow angle of incident fields because of the requirement for large antenna spacing. Polarization diversity is another way to achieve diversity gain while the antenna elements are colocated.

The propagation of horizontally and vertically polarized multipath components in a wireless channel is generally different since the reflection and diffraction processes depend on polarization. Even if the signal at the transmit antenna is in a signal polarized, the signal may be depolarized in the channel. The difference in propagation of the different polarized singles yields statistically independent fading. Polarization diversity can be then used at the receiver by using a dual-polarized antenna. Apparently, there are only two diversity paths in total.

In polarization diversity, the transmit signal can be generated in two ways. One method is to send the same signal in both vertical and horizontal polarization. In this way, the power should be divided in the two paths. Another realization is to transmit the signal in either vertically or horizontal polarized. In both cases, the signal can be received in both vertical and horizontal polarization because of the signal propagation environment. Let's consider the latter case that the signal is transmitted in horizontal polarization. In that case, since the fading of the received signals is independent, the average received signal strength in the two diversity branches is generally different.

Depending on the environment, the horizontal (i.e., copolarized) component is some 3–20 dB stronger than the vertical (cross-polarized) component. This has an important impact on the effectiveness of the diversity scheme. Research has been done on how to arrange the antenna to mitigate this problem.

In Andrews et al. (2001), the author claimed that by arranging three possible components of the E-field and three components of the H-field, a diversity order up to 6 may be achieved. However, full exploitation of that diversity order is generally prevented by the channel characteristics and practical considerations even though the diversity benefit of order 2 is of great value.

5.1.4 Frequency diversity

Frequency diversity transmits information on two or more carrier frequencies that are separated by more than the channel coherence bandwidth so that they will experience independent fades. If the channels are uncorrelated, the outage probability where deep fade occurs will be the product of outage probabilities of the individual branches.

Frequency diversity often applies in microwave line-of-sight links which convey information in multiple channels in frequency division multiplexing mode (FDM). In practice, deep fade occurs accidentally. Suppose there are N back up carrier frequencies labeled 1 to N, generally one carrier frequency is designed to be idle and can be available on stand-by basis to provide diversity switching for any one of the other N carrier frequencies to support an independent traffic.

This kind of diversity mode can also be applied in the channels in the orthogonal frequency division multiplexing (OFDM) transmission, where the subcarriers must be carefully designed so that the separation of adjacent subcarriers is bigger than the channel coherence bandwidth. We will discuss this issue in Chapter 7.

Sometimes spread spectrum is also regarded as providing frequency diversity since the channel gain varies across the bandwidth of the transmitted signal. However, this kind of frequency diversity does not mean transmitting the same information over different carrier frequencies. Spread spectrum with RAKE receiver provides independent fading paths of the information signal and thus it can be regarded as a kind of frequency diversity, and we will discuss it in Chapter 6 separately.

5.2 Channel coding over fading channel

By introducing redundancies, channel coding can *detect* or *correct* errors of a modulated signal occurred during transmission over wireless channel. According to its capability of detecting or correcting errors, the codes are classified into error detection codes and error correction codes (Cover et al. 2012). Channel coding may be designed for AWGN channels or specially for fading channels. In general, the codes designed for AWGN channels do not work well on fading channels since the long error bursts occur in deep fading cannot be corrected. Generally, the codes designed for fading channels are based on an AWGN channel code combined with interleaving. In this viewpoint, the main criterion of the code is not the coding gain, but the diversity in fading channel it provides. Other coding techniques such as unequal error protection codes and joint source can also work well in fading channels for their fading countermeasure features. Since channel coding may be covered by other courses, this chapter focuses on the issues related to channel coding over wireless channels.

5.2.1 Fundamentals of channel coding

The main task of error correction codes in a wireless system is to reduce the probability of bit or block error by introducing redundancies. Because of the redundant bits, the signal bandwidth will be increased if we want to keep the information rate constant, or the information rate will be decreased if we want to keep the signal bandwidth unchanged.

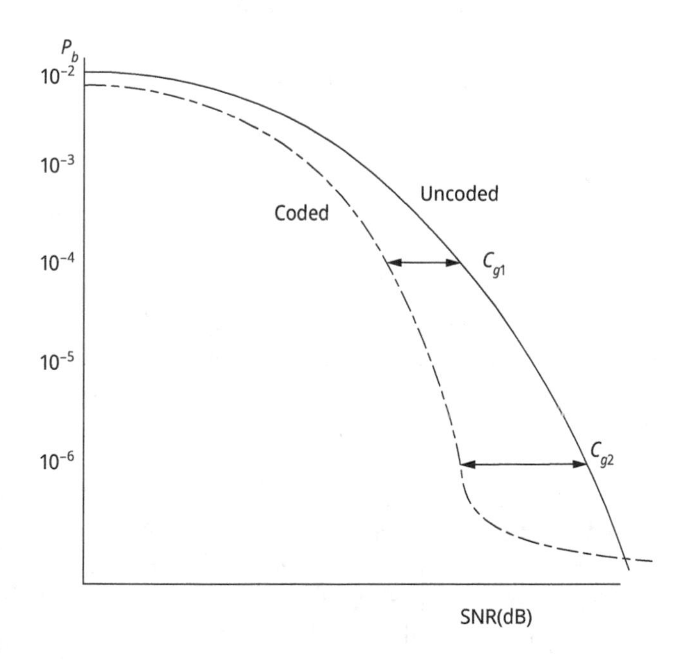

Fig. 5.4: Coding gain under AWGN channel.

According to the Shannon theory, the channel capacity in AWGN channel is

$$C = B\log_2(1+\gamma) = B\log_2\left(1+\frac{P}{N_0 B}\right) \tag{5.22}$$

Then, the capacity can be normalized by the bandwidth as

$$C/B = \log_2(1+\gamma) = \log_2\left(1+\frac{P}{N_0 B}\right) \tag{5.23}$$

where *C/B* is called the *bandwidth efficiency*. Apparently, the coding reduces the bandwidth efficiency in high SNR conditions, but might decreases the BER significantly at the low SNR conditions.

In designing a channel code, we want to obtain a positive coding gain. The coding gain is defined as the amount of the SNR that can be reduced under a certain coding technique for a given bit error probability P_b or block (frame) error probability P_{bl}. Figure 5.4 shows the code gain in AWGN channel. Apparently, the code gains at $P_b = 10^{-6}$ is greater than that at $P_b = 10^{-4}$ and the code gain at $P_b = 10^{-2}$ is negligible. The coding gain may not always be positive. If a code is designed for channels at high SNR, the channel gain may be negative at low SNRs for the redundancy bits are not sufficient to provide positive gain at low SNRs; thus the performance is degraded by the code.

There are two main classes of channel coding: *block coding* and *convolutional coding*. We will discuss the fundamental of them.

5.2.2 Block coding

5.2.2.1 Terms and notations

In block coding, the information source bits are grouped into blocks, and then a longer codeword which is actually transmitted in the channel is calculated from the information bits in the block.

For the convenience of description, a set of terms and notations are defined as follows:

(1) *Code rate:* Suppose the source information bits are grouped into blocks of K uncoded bits and each of the uncoded datablocks is encoded into a codeword of length N, then we define the ratio K/N the code rate R_c. For the unbinary codes, the basic information unit will be changed to the symbol from bit and the symbol alphabet of coded and undoded data is assumed the same.

(2) Binary codes: When the symbol alphabet is binary, using only "0" and "1," the code is called binary code. In Binary coding, "sum" means "modulo-2 sum," and "+" denotes a modulo-2 addition. In this chapter, we mainly discuss binary codes unless stated specifically.

(3) Weight of a code: The weight of a codeword is the number of nonzero elements in the codeword.

For example, the weight of codeword 11001101011 is 6.

(4) Distance of two codes: The distance of two codes is defined as the number of different elements in two codewords:

$$d(c_i, c_j) = \sum_{l=1}^{N} c_{i,l} \oplus c_{j,l} (\text{modulo} - q) \tag{5.24}$$

where c_i and c_j are two different codewords, d is the distance of the codeword and q is the number of possible values of c_i and c_j. For binary code, distance eq. (5.24) is refereed as *Hamming distance*.

And the minimum distance is defined as

$$d_{\min} = \min\{d(c_i, c_j)\} \tag{5.25}$$

(5) Linearity of a code: If the sum of any two codewords is also a valid codeword, the code is called linear code. The all-zero codeword must be contained in a linear code. All the codewords can be represented by a linear combination of basic codewords, which are called *generator words*.

(6) Systematic code: The systematic code is defined as the code in which the parity bits are appended to the end of the information bits. For a code of (n,k), the code word consists of k information bits which are the first k bits and $n-k$ appended at the end of the k bits and they are linear combination of the k information bits.

(7) Cyclic code: A cyclic code is one kind of linear code which satisfies the following property:

If $C = [c_{n-1}, c_{n-2}, \ldots, c_0]$ is a code word of a cyclic code, a cyclic shift of the code elements of C, that is, $[c_{n-2}, c_{n-3}, \ldots, c_0, c_{n-1}]$ is also a code word. Apparently, all cyclic shifts of C would be code words.

(8) Polynomial representation of a codeword: A codeword of length N can be represented by a polynomials of degree $\leq N-1$. In the polynomial, the nonzero coefficients correspond to the nonzero entries of the codevector; the variable x is a dummy variable. For example, a code word $x = [0\ 1\ 1\ 0\ 1\ 0\ 0\ 1]$ can be represented by

$$X_{(X)} = 0 \cdot x^7 + 1 \cdot x^6 + 1 \cdot x^5 + 0 \cdot x^4 + 1 \cdot x^3 + 0 \cdot x^2 + 0 \cdot x^1 + 1 \cdot x^0 \tag{5.26}$$

(9) Field and Galois field: A field F defines addition and multiplication for operating on elements, and it is closed under these operations (i.e., both the sum or product of two elements yield a valid element).

For addition operation, the identity element is called *zero* element and is denoted as 0. In multiplication operation, the identity element is called *unit* element and is denoted as 1. Multiplication is distributive over addition. That is, for any three elements a, b and c in F, $a(b + c) = ab + ac$ holds.

The addition inverse of a is denoted by $-a$, where $a+(-a) = 0$. The multiplicative inverse of a, denoted by a^{-1}, where $a \cdot a^{-1} = 1$.

A Galois field (GF) (p) is a finite field with p elements, where p is a prime integer.

The most important Galois field is GF (2). It consists of only two elements, that is, 0 and 1 and is the smallest finite field. Its addition and multiplication rules are as follows:

$$\begin{cases} 0+0 = 1+1 = 0 \\ 0+1 = 1+0 = 1 \\ 0\times 0 = 0\times 1 = 1\times 0 = 0 \\ 1\times 1 = 1 \end{cases} \tag{5.27}$$

(10) Irreducible polynomials and primitive polynomials: A polynomial of degree N is irreducible means that it not divisible by any polynomial of degree less than N and greater than 0. A primitive polynomial $g(x)$ of degree m is an irreducible polynomial such that the smallest integer N for which $g(x)$ divides $(x^N + 1)$ is $N = 2m - 1$.

5.2.2.2 Encoding

The straightforward encoding method can be considered as a table mapping. Map a K-valued information word to an associated N-valued codeword. The algorithm just checks the input information bits and reads out the corresponding codeword. Of course, this table checking algorithm is inefficient in storage; for linear systematic code, this process can be described by the help of generator matrix.

Suppose **C** denotes the N-dimensional codevector, **U** denotes the K-dimensional information vector and $\boldsymbol{G}$ is the $K \times N$-dimensional generator matrix. Furthermore, we use row vectors to represent codewords, and the encoding is represented by

$$\boldsymbol{C} = \boldsymbol{UG} \tag{5.28}$$

For a systematic code, the $K \times N$-dimensional generator matrix has the following characteristics:

1) the leftmost K columns are a $K \times K$ identity matrix;
2) the right $N - K$ columns are the parity check bits.

Then G can be written as

$$G = [I_k|P] = \begin{bmatrix} 1 & 0 & \dots & 0 & | & p_{11} & p_{12} & \dots & p_{1,n-k} \\ 0 & 1 & \dots & 0 & | & p_{21} & p_{22} & \dots & p_{2,n-k} \\ \vdots & \vdots & & \vdots & | & \vdots & \vdots & & \vdots \\ 0 & 0 & \dots & 1 & | & p_{k1} & p_{k2} & \dots & p_{k,n-k} \end{bmatrix} \tag{5.29}$$

The first K bits of $\boldsymbol{C}$ in eq. (5.28) are identical to $\boldsymbol{U}$, and that a vector–matrix product is obtained by premultiplying the matrix with a row vector.

Then a codeword $\boldsymbol{x}$ of a symmetric encoder is generated as follows:

$$\mathbf{x} = \boldsymbol{uG} = \boldsymbol{U}_i[\boldsymbol{I}_k|\boldsymbol{P}] = \left[u_{i1}, \dots, u_{ik}, p_1, \dots, p_{(n-k)}\right] \tag{5.30}$$

where the first k bits in C_i are the original information bits and the last $(n-k)$ bits appended to the original bits are the parity bits obtained from the information bits as

$$p_j = u_{i1}p_{1j} + \dots + u_{ik}p_{kj},\ j = 1, \dots, n-k. \tag{5.31}$$

We can use an example to show the encoding process.

Example 5.1. Consider a (7,4) code, the generator matrix is

$$\boldsymbol{G} = \begin{bmatrix} 1 & 0 & 0 & 0 & | & 1 & 1 & 0 \\ 0 & 1 & 0 & 0 & | & 1 & 0 & 1 \\ 0 & 0 & 1 & 0 & | & 1 & 0 & 1 \\ 0 & 0 & 0 & 1 & | & 0 & 1 & 0 \end{bmatrix}$$

We encode the source word [1011]. By using eq. (5.30) we can compute the codeword

$$\boldsymbol{x} = \boldsymbol{uG} = [1\ 0\ 1\ 1]\begin{bmatrix} 1 & 0 & 0 & 0 & | & 1 & 1 & 0 \\ 0 & 1 & 0 & 0 & | & 1 & 0 & 1 \\ 0 & 0 & 1 & 0 & | & 1 & 0 & 1 \\ 0 & 0 & 0 & 1 & | & 0 & 1 & 0 \end{bmatrix} = [1\ 0\ 1\ 1\ 0\ 0\ 1] \tag{5.32}$$

From the generator matrix G and eq. (5.32), we can find that the first k bits of the codeword are identical to the original information bits so it is not necessary to calculate them and only the reminder $(n-k)$ bits are need to be calculated. The P in the matrix is

$$\boldsymbol{P} = \begin{bmatrix} 1 & 1 & 0 \\ 1 & 0 & 1 \\ 1 & 0 & 1 \\ 0 & 1 & 0 \end{bmatrix} \tag{5.33}$$

By using eq. (5.31), we can calculate the corresponding bits that appended to the original information bits. Equation (5.31) shows that the first parity bit in the codeword is $p_1 = u_{i1}P_{11} + u_{i2}P_{21} + u_{i3}P_{31} + u_{i4}P_{41} = u_{i1} + u_{i2}$. By a similar way, the second parity bit p_2 the third parity bit is p_3 can be found by $p_2 = u_{i1}P_{12} + u_{i2}P_{22} + u_{i3}P_{32} + u_{i4}P_{42} = u_{i1} + u_{i4}$ and $p_3 = u_{i1}P_{13} + u_{i2}P_{23} + u_{i3}P_{33} + u_{i3}P_{43} = u_{i2} + u_{i3}$. Shift registers may be used to generate the parity bits, which are shown in Fig. 5.5. The output codeword is then $[u_{i1}\ u_{i2}\ u_{i3}\ u_{i4}\ p_1\ p_2\ p_3]$, where the first four bits are output when the switch is in the down position and the parity bits are the output when switch is in the up position.

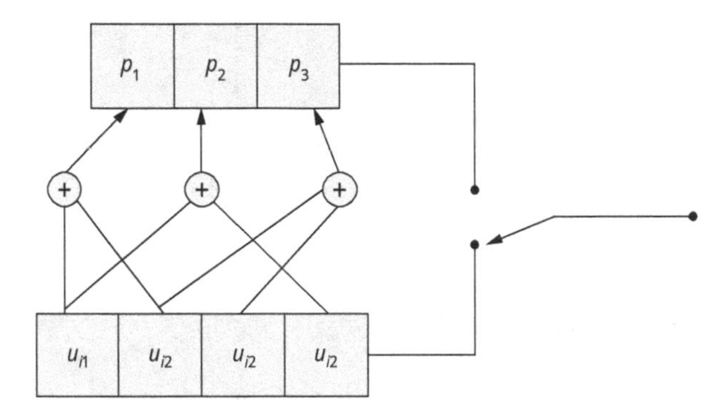

Fig. 5.5: Shift register implementation structure for (7,4) binary code.

5.2.2.3 Decoding

In the decoding procedure, a received codeword is first checked if it is a valid codeword. For an invalid codeword, it will be checked if these errors are correctable and then these errors may be corrected and a corresponding valid codeword is obtained. After that the transmitted information will be estimated from the codeword.

By multiplying a codeword with a parity check matrix H, an $N-K$-dimensional *syndrome vector* s_{synd} is obtained. If s_{synd} has all-zero entries, the received codeword is valid.

Since a valid codeword times H yields an all-zero s_{synd}, $\boldsymbol{H}\,\boldsymbol{G}^T = \boldsymbol{0}$ should be always hold. Substitute $\boldsymbol{G} = [\boldsymbol{I}|\boldsymbol{P}]$ which is expressed in eq. (5.29) into $\boldsymbol{H}\cdot\boldsymbol{G}^T = \boldsymbol{0}$, we get

$$\boldsymbol{H}\cdot\boldsymbol{G}^T = (\boldsymbol{H}_1\boldsymbol{H}_2)(\boldsymbol{I}\ \boldsymbol{P})^{\boldsymbol{T}} = (\boldsymbol{H}_1\boldsymbol{H}_2)\begin{pmatrix}\boldsymbol{I}\\ \boldsymbol{P}^T\end{pmatrix} = (\boldsymbol{H}_1 + \boldsymbol{H}_2\boldsymbol{P}^T) = \boldsymbol{0} \tag{5.34}$$

The equation will hold if we choose $H_2 = I$ and $H_1 = -P^T$ since the subtraction of two identical matrices is the all-zero matrix. Apparently, there are might several different parities check matrices corresponding to one generator matrix, and the method in eq. (5.34) is just one solution.

Example 5.2. Consider the generator matrix in Example 5.1,

$$G = \begin{bmatrix} 1 & 0 & 0 & 0 & | & 1 & 1 & 0 \\ 0 & 1 & 0 & 0 & | & 1 & 0 & 1 \\ 0 & 0 & 1 & 0 & | & 1 & 0 & 1 \\ 0 & 0 & 0 & 1 & | & 0 & 1 & 0 \end{bmatrix}$$

$$P = \begin{bmatrix} 1 & 1 & 0 \\ 1 & 0 & 1 \\ 1 & 0 & 1 \\ 0 & 1 & 0 \end{bmatrix} P^T = \begin{bmatrix} 1 & 1 & 1 & 0 \\ 1 & 0 & 0 & 1 \\ 0 & 1 & 1 & 0 \end{bmatrix}$$

Since in modulo 2 arithmetic, $-P^T = P^T$,

$$H = (H_1 \; H_2) = \left(-P^T I\right) = \begin{bmatrix} 1 & 1 & 1 & 0 & 1 & 0 & 0 \\ 1 & 0 & 0 & 1 & 0 & 1 & 0 \\ 0 & 1 & 1 & 0 & 0 & 0 & 1 \end{bmatrix}$$

Assume the received codeword $\widehat{x} = x + \varepsilon$, where x is a valid codeword and ε is an error word. The syndrome s_{synd} will be determined by the error word, that is

$$s_{\text{synd}} = \widehat{x}H^T = (x + \varepsilon)H^T = xH^T + \varepsilon H^T = 0 + \varepsilon H^T = \varepsilon H^T \tag{5.35}$$

A nonzero syndrome corresponding an error vector ($\varepsilon \neq 0$). Equation (5.35) corresponds to $n - k$ equations in n unknowns, there are $2k$ possible error patterns that can produce a given syndrome $s_{\text{synd.}}$ Finding the "correct" ε and subtracting it from the received codeword, we will get the correct codeword. It is easy to understand that one syndrome may correspond to a number of different error words ε and we call these error words a *coset*. The key task of decoding is to find the most probable ε corresponding to the current syndrome. We assume that the probability that a bit arrives correctly is higher than the probability that a bit arrives with an error. This assumption is correct when the SNR of the channel is "reasonable." Therefore, we generally pick the ε that has the minimum weight in a given *coset*.

5.2.3 Convolutional codes

5.2.3.1 Encoding of convolutional codes

Convolutional codes do not group the source data blocks before encoding, but rather generate coded symbols by passing the information bits through a linear finite-state shift register as shown in Fig. 5.6.

Let's use (n, k, K) to denote a convolutional code, where k is the number of bits input to the encoder each time, n is the length of codeword corresponding to input k tuples, K is the stages of the k tuples of the encoder which is called constraint length.

Each time, the convolutional encoder encodes k tuples into n tuples, and k and n is generally very small, so the coding delay is very short. The output of an encoder is determined not only by the current input, but also by the previous $K-1$ input tuples.

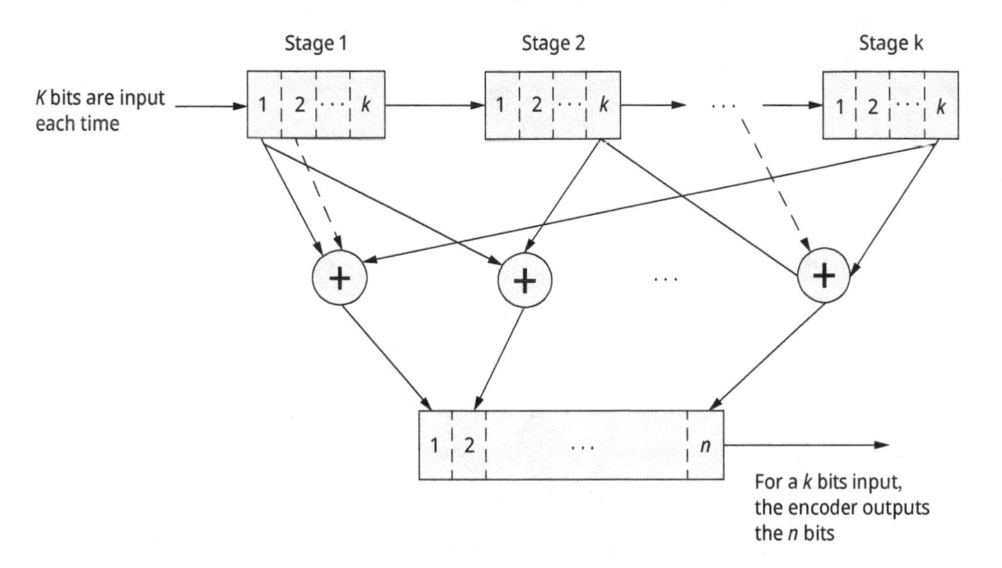

Fig. 5.6: A general diagram for the convolutional encoding principle.

Example 5.3. An encoder for a convolutional code of $(n, k, K) = (3, 1, 3)$ is shown in Fig. 5.7.

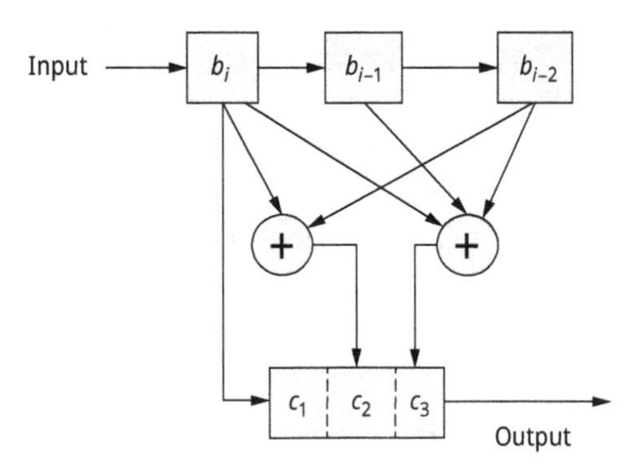

Fig. 5.7: A convolutional encoder for (3, 1, 3).

When one bit is input, there are three corresponding output bits:

$$c_1 = b_i$$
$$c_2 = b_i \oplus b_{i-2}$$
$$c_3 = b_i \oplus b_{i-1} \oplus b_{i-2}$$

Therefore, the code rate is 1/3.

The output of the encoder depends on the current input in b_i and the pervious inputs b_{i-1} and b_{i-2} so the bits in register b_{i-1} and b_{i-2} can be regarded as the state of the encoder. There are $2^{K-1} = 4$ states in this example.

State	b_{i-1}	b_{i-2}
S_1	0	0
S_2	1	0
S_3	0	1
S_4	1	1

In the above example, for the same input, the output of the encoder may be different because of the difference of the states. In practice, we generally assume the encoder start from all-zero state. After all the necessary information bits are put into the registers and the required bits are obtained, we will use zeros as the input to the register until the last information bit are pushed out of the register, and all the registers become all-zero state again.

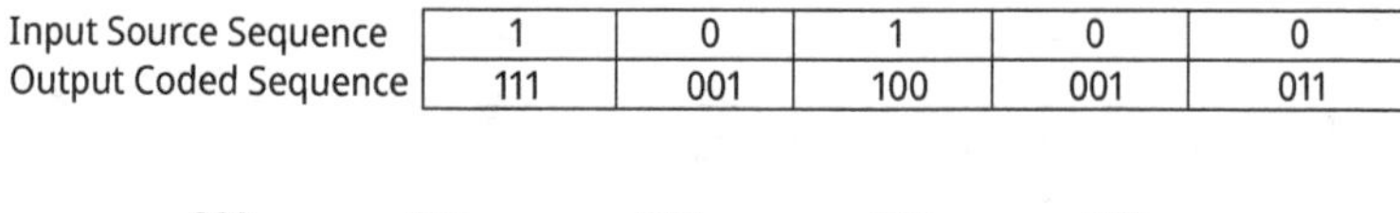

Input Source Sequence	1	0	1	0	0
Output Coded Sequence	111	001	100	001	011

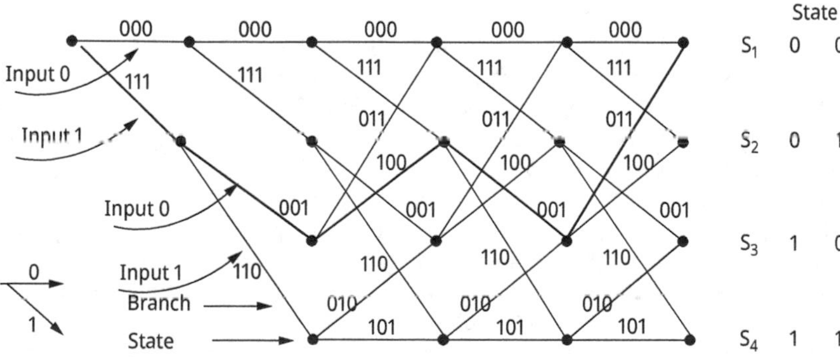

Fig. 5.8: Trellis description for Fig. 5.7.

In describing the encoding and decoding, the trellis diagram is often used. The trellis shows the input, state and corresponding output. Figure 5.8 gives trellis of the encoder. There are two branches originate from each state: the upper branch is for input bit 0 and the lower for bit 1. Apparently, each time there will be only two possible state transitions for one state, and this redundant feature can be used in the decoding process. The trellis will be repeated after a certain period. Therefore, a limited length of trellis is enough to represent the encoding of input sequence with an infinitely length.

5.2.3.2 Decoding of convolutional codes

If the received bit sequence is identical to the source bit sequence, the decoder can trace the trellis directly to find the original source bit sequence. However, if there are errors in the received bit sequence because of noise or interference, we have to estimate the sequence from the received sequence. We will introduce one of these estimation algorithms, the Viterbi algorithm to show the principle.

The Viterbi algorithm was proposed by Viterbi in 1967, which is an algorithm that has a low complexity because it systematically removes paths that cannot achieve the highest path metric. The strategy is to check all the partial path metrics of the paths that entering one node and keep the best one. For the possible paths *leaving* one certain node in the trellis are the same for all paths that *enters* the node, if the complete trellis path includes the highest metric path (or smallest cost path), it must be exactly coinciding with the path that has the highest partial path metric up to the considered node.

Example 5.4. Figure 5.9 shows the decoding procedure. Figure 5.9(a) shows the basic structure of the trellis. The bit sequence received and to be decoded is given in Fig. 5.9(b). The metrics we used are the Hamming distances between the received sequence and the theoretically possible bit sequences in the trellis. For example, in Fig. 5.9(c), for passing the first bit sequence 101, the Hamming distance of it with the theoretically possible bit sequence of upper branch 000 is 2, and that of it with the theoretically possible bit sequence of lower branch 111 is 1. In each state at a time point $s_i(j)$, where i denotes the state and j denote the time, we check all possible branches from the previous time point. For example, at $s_1(3)$ at the third sequence time at Fig. 5.9(c), there are two possible branches entering state $s_1(3)$, one is $s_1(0)$—>000 —> $s_1(1)$—>000—> $s_1(2)$—>000—> $s_1(3)$ with partial distance is 3, the other is $s_1(0)$—>000—> $s_2(1)$—>000 —> $s_3(2)$—>000—> $s_1(3)$ with partial distance is 5 which will be discarded in the next step. This procedure is repeated till the end shown in Fig. 5.9(d), where we can find the total branch distances at $s_1(5)$, $s_2(5)$, $s_3(5)$ and $s_4(5)$. By trace back the path, we can find that the output of encoder should be 110,001,100,001,011 and the decoding result is 1,0,1,0,0.

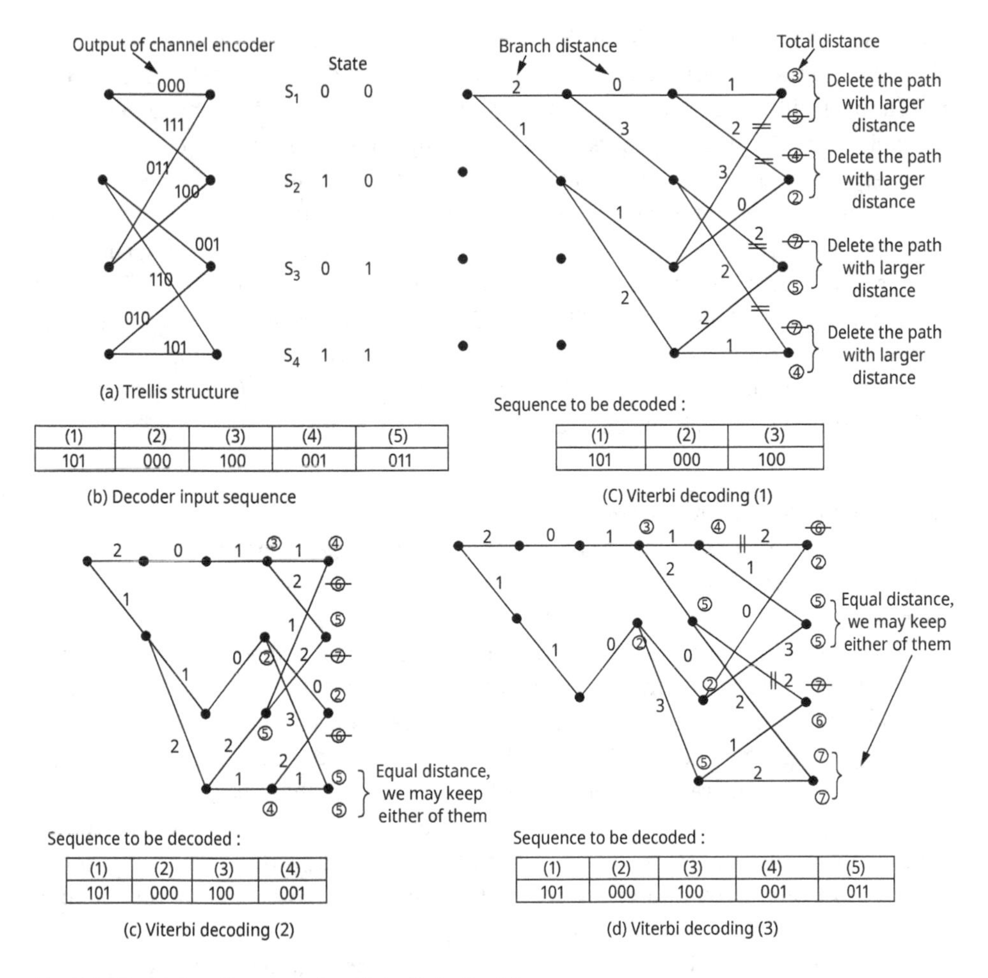

Fig. 5.9: Viterbi decoding example with trellis in Fig. 5.8.

5.2.4 Coding and interleaving for fading channels

We have discussed two typical channel coding methods, block coding and convolutional coding, which are designed for AWGN channel. From Chapters 2 and 3, we know that the wireless channel is in fading in most cases, where the errors of the digital modulations tend to occur in bursts during the deep fading period. The error bursts generally cannot be corrected by the codes designed for AWGN channels. Therefore, the coding may exhibit worse performance and get a negative coding gain.

In order to correct the error bursts, coding can be combined with interleaving to randomize and spread the error bits to many codewords so that there are only a few simultaneous errors in one codeword which is within the correcting capability of the codeword (Ramsey 1970). The size of the interleaver should be big enough so that the fading across a received codeword is independent, and this will cause large delay when the channel coherence time is large or equivalently the channel is in slow fading. Channel coding combined with interleaving can be regarded as time diversity. In this chapter, we'll mainly discuss the interleaving combined with block coding and the interleaving combined with convolutional coding. Although turbo codes have an interleaver inherent to the code design, the design of interleaver is mainly for AWGN channel. There are also other coding techniques that can work under fading channels, such as Low Density Parity Check (LDPC) Codes, Trellis Coded Modulation (TCM), Bit Interleaved Coded Modulation (BICM) and Unequal error protection codes. We'll not discuss these techniques in this chapter, and interested readers can find more information in the related references (Gallagher 1961, Ayanoglu et al. 1987, Ungerboeck 1982, Goldsmith et al. 1998, Vuceti et al. 2000).

5.2.4.1 Block coding with interleaving

Generally, the deep fading in the channel will cause a large number of bit errors that cannot be corrected by a normal code. We first discuss the structure of block interleaving, which can be shown in Fig. 5.10.

In the figure, we assume each row is a codeword. The codewords are filled into the interleaver matrix by row. After the matrix is full, the data is read out by columns. Therefore, two symbols (or bits, depends on the code) adjacent in the channel are separated by $n - 1$ other symbols, where n is the length of a codeword. Two symbols adjacent in the codeword are separated by $D - 1$ symbols in the channel.

In the figure, if $DT_S > T_c$, where T_c is the channel coherence time, the interleaver is called a deep interleaver so that an error burst caused by a deep fading will introduce at most one error in a codeword that can be easily corrected by the channel decoding.

The size of the interleaver can be determined by the channel coherence time, codeword size and the error correcting ability of the codewords.

It should be mentioned that an interleaver can reduce the mean BER only in combination with a channel code. Without a channel code, the interleaver can also break up error bursts, but it does not yield a lower mean BER. However, breaking up the errors may be desirable in some services.

Another constrain in using an interleaver is the delay it caused. In some scenario, a maximum delay is required which may limit the use or effectiveness of the interleaver.

5.2.4.2 Convolutional coding with interleaving

Similar to the block coding, convolutional codes which are designed for AWGN channel also do not work effectively in the fading environments where the deep fading occurs. Interleaver may applied to the convolutional codes to spread out the errors occurred in deep fading period. Figure 5.11 demonstrates such an example.

There are N parallel buffers with increasing size from 0 to N–1 between the output of the encoder and the input of the modulator. The output of encoder is multiplexed to the buffers, and the input to the modulator is multiplexed from the buffer. Therefore, two adjacent output symbols from the encoder would be separated by N–1 other symbols in the modulator and channel. Similarly, there are N parallel buffers at receiver, but the sizes of the buffers are in a decreasing order from N–1 to 0. The output from the demodulator is multiplexed to the buffers and the input to the decoder is multiplexed from the buffers.

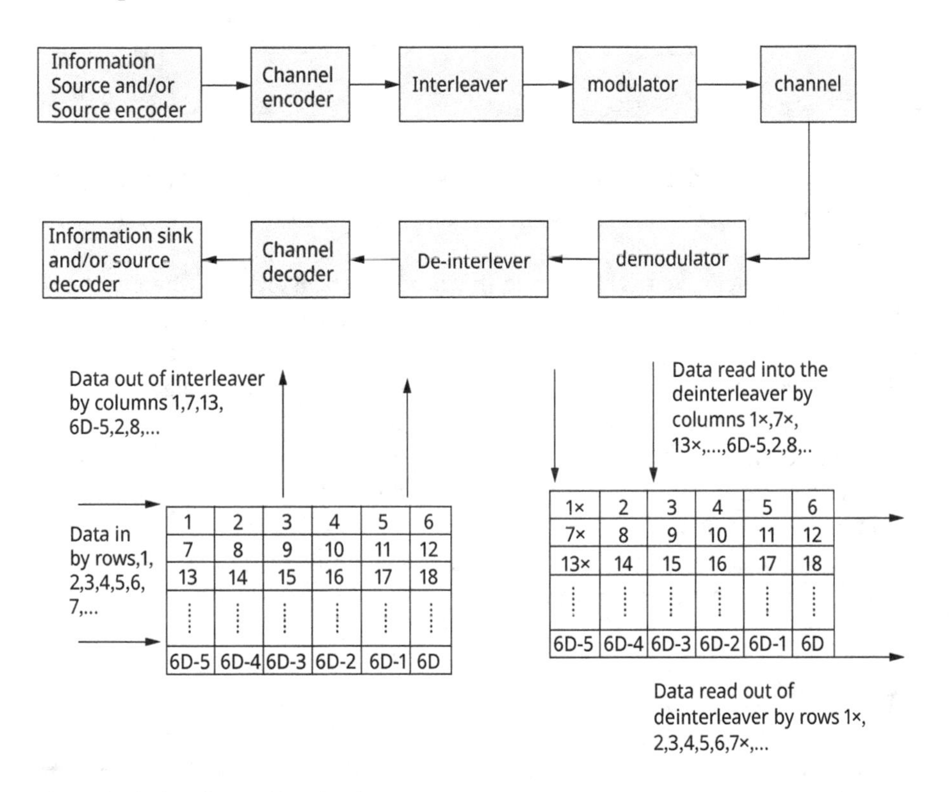

Fig. 5.10: Block coding and interleaving.

5.3 Equalization

Multipath components in wireless channels may have different runtimes from the transmitter (TX) to the receiver (RX), and this will introduce delay spread and lead to ISI. From Chapter 4 we know that, ISI can cause signal distortion and will yield an irreducible error floor when the modulation symbol duration is on the same order as the channel delay spread. Several signal processing methods can be used to counter-

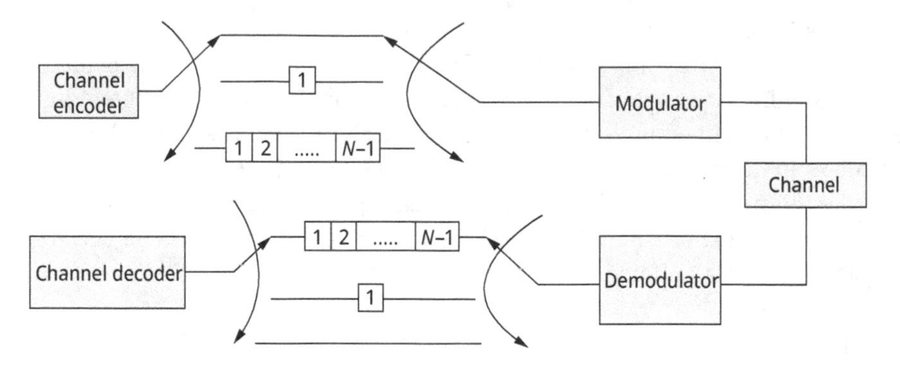

Fig. 5.11: Convolutional coding with interleaving.

act ISI and equalization denotes the signal processing technique to alleviate ISI at the receiver. Other signal processing methods used at the transmitter will be treated in the later chapters.

When the channel's rms delay spread σt_m is at the same order of the inverse of signal bandwidth $1/B_u$, measures should be taken to overcome ISI. For example, equalization is generally required in the standard cellular phone systems because $1/B_u \approx \sigma t_m$ whereas it is not necessary in cordless phones since the delay spread in indoor environment is small comparing to the symbol period of the fairly low data rate of the voice. In fact, equalization is a typical component in high data rate wireless applications since they are more sensitive to delay spread, otherwise, alternative ISI mitigation methods must be used.

In this section, we will first discuss the time domain equalization and then describe the frequency domain equalization.

5.3.1 Time domain equalization

Time domain equalization can be classified into two main categories: linear and nonlinear. The linear methods are typically simple and easy to understand, while the nonlinear equalizers provide better performance and therefore are used in most applications. Decision-feedback equalization (DFE) is the most common nonlinear method, but it has a poor performance because of error propagation with decoding errors existed. The method of maximum likelihood sequence estimation equalization is optimal but is impractical in complexity, so it is generally used as an upper bound on performance to evaluate other algorithms. Figure 5.12 gives the outline of different equalization methods. We will discuss some of them in this section.

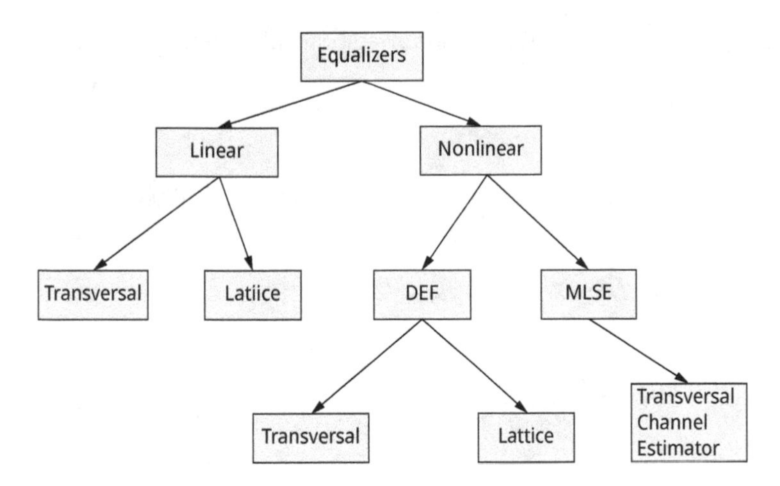

Fig. 5.12: Classification of equalizations.

5.3.1.1 Linear equalizer

A communication system with adaptive equalizer is shown in Fig. 5.13. We assume the equalizer in the figure is implemented via a linear $N = 2L + 1$ tap transversal filter:

$$H_{\text{eq}}(z) = \sum_{i=-L}^{L} w_i z^{-i} \tag{5.36}$$

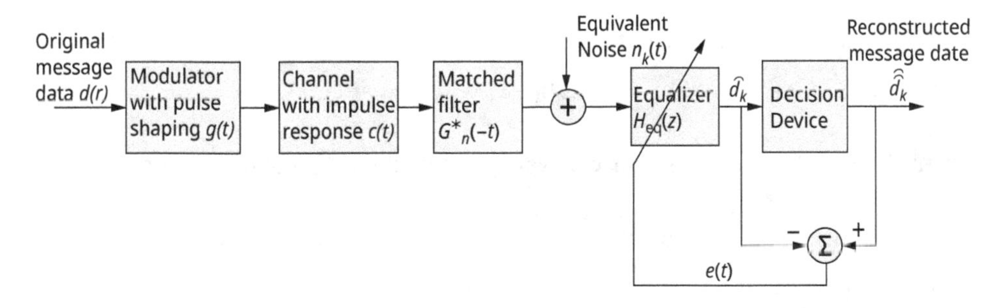

Fig. 5.13: Diagram of a communication system with an adaptive equalizer.

As we know, for causal linear filters, the weights $w_i = 0$, $i < 0$. If the size N of the equalizer is given, the tap weights w_i, $i = -L, -L+1, \ldots, 0, \ldots, L-1, L$ must be specified for a specified channel frequency response. The weights will vary as the channel changes. We generally optimize the equalizer coefficients by minimizing the probability of error (or outage probability) which is the performance metric we discussed in Chapter 4. Unfortunately, the probability of error (or outage probability) cannot be directly optimized subject to the equalizer coefficients; however indirect optimization can be used to balance between ISI mitigation and noise enhancement. We will discuss two types of linear equalizers: Zero Forcing (ZF) equalizer and the Minimum Mean Square Error (MMSE) equalizer.

(1) ZF equalizer

In Fig. 5.13, suppose $f(t)$ is the combined impulse response of the transmitter, channel and matched filter:

$$f(t) = g(t) * c(t) * g_m^*(-t) \tag{5.37}$$

Let's define

$$h(t) = g(t) * c(t) \tag{5.38}$$

We can rewrite $f(t)$ as

$$f(t) = h(t) * g_m^*(-t) \tag{5.39}$$

Then, the input to the equalizer $y(t)$ is

$$y(t) = d(t) * f(t) + n_g(t) = \sum_{k=-\infty}^{\infty} d_k f(t - kT) + n_g(t) \tag{5.40}$$

where T is the symbol time. Then the input sample $y_n = y[n] = y[nT]$ is

$$y_n = \sum_{k=-\infty}^{\infty} d_k f(nT_s - kT_s) + n_g(nT_s) \tag{5.41}$$

In Z-domain, it can be written as

$$Y(z) = D(z)F(z) + N_g(z) \tag{5.42}$$

where $N_g(Z)$ is the equivalent noise power spectrum after the matched filter:

$$N_g(z) = N_0 |G_m^*(1/z^*)|^2 \tag{5.43}$$

and $F(Z)$ can be written as

$$F(z) = H(z) G_m^*(1/z^*) = \sum_n f(nT_s) z^{-n} \tag{5.44}$$

The function of zero-forcing equalizer is to cancel all ISI effect introduced in the combined response $f(t)$. It can be seen from eq. (5.42)

$$H_{ZF}(z) = 1/F(z) \tag{5.45}$$

At the same time, the noise spectrum after the equalizer would be

$$N(z) = N_g(z)|H_{ZF}(z)|^2$$
$$= \frac{N_g(z)}{|F(z)|^2} = \frac{N_0|G_m^*(1/z^*)|^2}{|F(z)|^2} = \frac{N_0|G_m^*(1/z^*)|^2}{|H(z)|^2|G_m^*(1/z^*)|^2} = \frac{N_0}{|H(z)|^2} \quad (5.46)$$

Please notice that the channel $H(z)$ in the denominator of the above equation may be severely attenuated in certain bandwidth on frequency-selective-channels, and this will enhance the noise significantly. An equalizer needs balance the ISI attenuation and the noise enhancement, and it would be considered in the MMSE equalizer.

(2) MMSE equalizer

As discussed above, the MMSE is an indirect optimization method that balance between ISI mitigation and the prevention of noise enhancement by minimizing the expected mean-squared error between the symbol detected at the equalizer output and the transmitted symbol, and this kind of balance makes the MMSE equalizers outperforms ZF equalizers in BER performance.

Suppose the equalizer is defined in eq. (5.36) and the input signal to the equalizer can be represented as a vector $\boldsymbol{y}_k$

$$\boldsymbol{y}_k = [y_k \; y_{k-1} \; y_{k-2} \ldots y_{k-L}]^T \quad (5.47)$$

The output of the equalizer should be

$$\widehat{d}_k = \sum_{l=0}^{L} w_{lk} y_{k-l} \quad (5.48)$$

where the weight vector can be written as

$$\boldsymbol{w}_k = [w_{0k} \; w_{1k} \; w_{2k} \; w_{3k} \ldots w_{Lk}]^T \quad (5.49)$$

By using vector notation, eq. (5.48) can be rewritten as

$$\widehat{d}_k = \boldsymbol{y}_k{}^T \boldsymbol{w}_k = \boldsymbol{w}_k{}^T \boldsymbol{y}_k \quad (5.50)$$

In the training period, the transmitter sends a known sequence to estimate these coefficients of the equalizer; therefore the error can be calculated from the difference between the output $\widehat{d}_k$ and the desired one $d_k = x_k$,

$$e_k = \widehat{d}_k - d_k = \widehat{d} - x_k \quad (5.51)$$

Then, we can rewrite *ek* from eq. (5.50)

$$e_k = x_k - \widehat{d}_k = x_k - \boldsymbol{y}_k{}^T \boldsymbol{w}_k = x_k - \boldsymbol{w}_k{}^T \boldsymbol{y}_k \quad (5.52)$$

Then the squared error at time k is

$$|e_k|^2 = x_k{}^2 - \boldsymbol{w}_k{}^T \boldsymbol{y}_k \boldsymbol{y}_k{}^T \boldsymbol{w}_k - 2x_k \boldsymbol{y}_k{}^T \boldsymbol{w}_k \tag{5.53}$$

And the mean-squared error is obtained by taking the expectation of it.

$$E[|e_k|^2] = E[x_k{}^2] - \boldsymbol{w}_k{}^T E[\boldsymbol{y}_k \boldsymbol{y}_k{}^T] \boldsymbol{w}_k - 2E[x_k \boldsymbol{y}_k{}^T] \boldsymbol{w}_k \tag{5.54}$$

In practice, the time average can be used to substitute the expectation. Since we assume that the weights are unchanged in the time of interest, the filter weights wk are not included in the expectation operation in eq. (5.54). We define the cross-correlation vector $\boldsymbol{p}$ between the desired response and the input signal as

$$p = E[x_k \boldsymbol{y}_k] = E[x_k y_k \quad x_k y_{k-1} \quad x_k y_{k-2} \ldots \quad x_k y_{k-N}] \tag{5.55}$$

And the $(N+1) \times (N+1)$ input correlation matrix, sometimes called the input covariance matrix, is defined as

$$\boldsymbol{R} = E[\boldsymbol{y}_k \boldsymbol{y}_k^*] = E \begin{bmatrix} y_k{}^2 & y_k y_{k-1} & y_k y_{k-2} & \cdots & y_k y_{k-N} \\ y_{k-1} y_k & y_{k-1}^2 & y_{k-1} y_{k-2} & \cdots & y_{k-1} y_{k-N} \\ \vdots & \vdots & \vdots & \cdots & \vdots \\ y_{k-N} y_k & y_{k-N} y_{k-1} & y_{k-N} y_{k-2} & \cdots & y_{k-N}^2 \end{bmatrix} \tag{5.56}$$

The mean square values of each sample are contained in the major diagonal of $\mathbf{R}$. By using eqs. (5.55) and (5.56), the mean square error in eq. (5.54) can be rewritten as

$$\xi = E[x^2] + w^T \mathbf{R} w - 2p^T w \tag{5.57}$$

Minimizing the mean square error in terms of the weight vector wk, the optimal coefficients for the equalizer can be found and updated. In this way, the receiver will get a flat spectral response will a minimal ISI.

It should be mentioned that, in the above method, there must be known sequence used as the training sequence. A dedicated training sequence may be designed in the frame structure such as the pilots shown in Fig. 5.14.

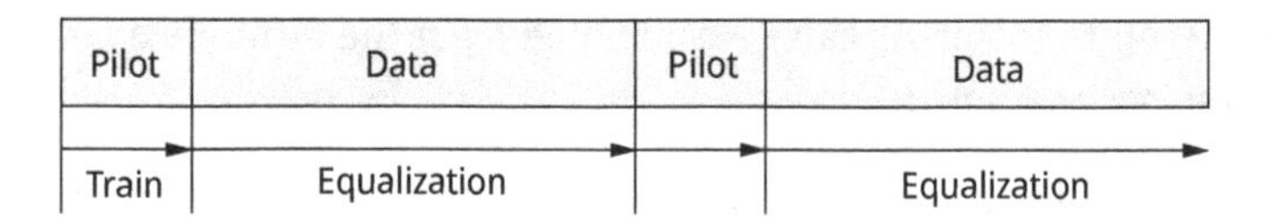

Fig. 5.14: Frame structure for MMSE equalizer.

5.3.1.2 Decision-feedback equalizer

Decision-feedback equalizer is one kind of nonlinear equalizer, and it consists of two filters: a feedforward filter $F(z)$ and a feedback filter $B(z)$, as shown in Fig. 5.15.

The input to the feedforward filter $F(z)$ is the received sequence which is similar to linear equalizer, while input to the feedback filter $B(z)$ is previously detected sequence.

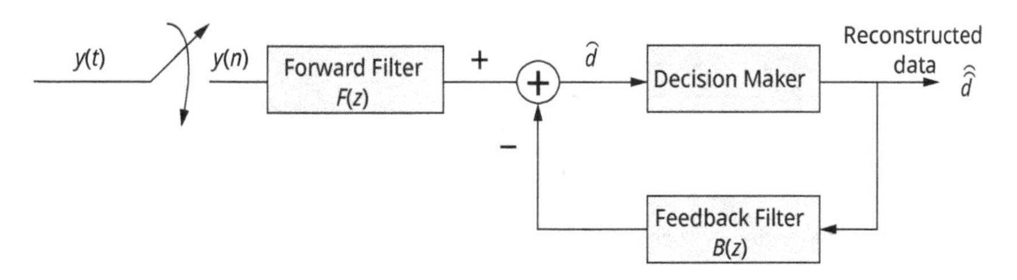

Fig. 5.15: Decision-feedback equalizer.

The previously detected symbols $\{\widehat{\hat{d}}_n\}$ is passed to the Feedback filter $B(z)$ that approximates the convolution of baseband channel and the forward filter, and the output can be used to estimate the ISI contribution of the previous symbols. As shown in Fig. 5.15, the value subtracted the ISI contribution from the output of the forward filter is then put into the decision maker. The feedback filter $B(z)$ must be strictly causal for it is in a feedback loop, or else the system will be unstable. Because the feedback filter of the DFE estimates the channel frequency response rather than its inverse, it does not suffer from noise enhancement. Therefore, the DFEs generally outperform the linear equalizers under channels with deep spectral nulls.

5.3.2 Frequency domain equalizer

We have already discussed time-domain equalizers which are popular in wireless applications to overcome ISI. However, frequency domain equalizers (FDE) are required for OFDMA-based systems which are used in high-speed communications. Also as the data rate increases the complexity of time domain equalizer grows since the channel delay spread spans more symbols. Another scenario is that the equalizer complexity grows in higher-order modulation. In these situations, frequency domain signal processing is expected. In this section, we will discuss FDE in single carrier (SC) wireless systems.

Figure 5.16 shows a general SC-FDE-based receiver with multiple antennas. By using discrete Fourier transform (DFT), the received signal is transformed to the frequency domain. Then, each branch is applied an equalizer weight by multiplication. After that, the signals are combined together and the transformed back into the time domain by an inverse DFT (I-DFT) operation.

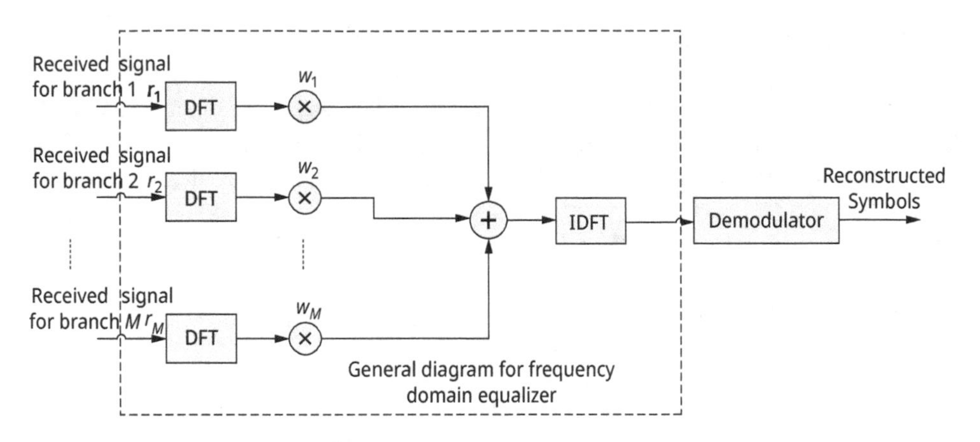

Fig. 5.16: Diagram for single-carrier receiver with FDE.

On determining the size of the DFT, we need consider the size should be enough to represent the inverse of the channel response in the time domain. The weights used to combine the branches are determined by the MMSE criteria:

$$w_j(k) = \frac{\left(\frac{\sigma_d}{\sigma_j}\right)^2 \cdot H_j^*(k)}{\sum_{i=1}^{M} \left(\frac{\sigma_d}{\sigma_i}\right)^2 \cdot |H_i(k)|^2 + 1} (j = 1, 2, \ldots, M) \tag{5.58}$$

where σ_d^2 above is the desired signal power, and σ_j^2 denotes the noise power on the jth antenna. Let's now assume a simple case where $M = 2$, the weights can be found as

$$w_1(k) = \frac{H_1^*(k)}{|H_1(k)|^2 + \left(\frac{\sigma_1}{\sigma_2}\right)^2 \cdot |H_2(k)|^2 + \left(\frac{\sigma_1}{\sigma_d}\right)^2} \tag{5.59}$$

$$w_2(k) = \frac{H_2^*(k)}{\left(\frac{\sigma_2}{\sigma_1}\right)^2 \cdot |H_1(k)|^2 + |H_2(k)|^2 + \left(\frac{\sigma_2}{\sigma_d}\right)^2} \tag{5.60}$$

Of course, we can assume that the power of noise is the same for the two receive antennas; in such a way the denominator becomes the same for the both weights.

Problems

5.1 Figure out different methods to implement independent paths for diversity.

5.2 Describe the main idea of the combining techniques for receiver diversity. What are the main advantages and disadvantages of these combining techniques?

5.3 Assume each of the equal branches has an equal SNR of 7 dB, what is the outage probability of BPSK modulation at $P_s = 10^{-3}$ for a Rayleigh fading channel with SC diversity for $M = 1$ (no diversity), $M = 2$ and $M = 3$.

5.4 Find the outage probability of QPSK modulation at $P_b = 10^{-3}$ under MRC and EGC for two-branch diversity, assume there is i.i.d. Rayleigh fading on each branch and average branch SNR $\bar{\gamma} = 10$ dB.

5.5 Suppose the generator matrix of a (7,4) code is given below:

$$G = \begin{bmatrix} 0 & 1 & 0 & 1 & 1 & 0 & 0 \\ 1 & 0 & 1 & 0 & 1 & 0 & 0 \\ 0 & 1 & 1 & 0 & 0 & 1 & 0 \\ 1 & 1 & 0 & 0 & 0 & 0 & 1 \end{bmatrix}$$

(a) Find the codewords of the code.
(b) What is the minimum distance of the code?
(c) Give the parity check matrix of the code.

5.6 Consider a channel with coherence time $T_c = 10$ ms and a coded bit rate of $R_s =$ 50,000 kilosymbols per second. For a (7,4) block code, find the length of the interleaver matrix so that the fade dip will only affect a single bit of a code word. Discuss the average delay of this coding scheme.

5.7 Figure out the different types of equalizers and discuss the advantages and disadvantages of different types.

Chapter 6
Spread spectrum systems and spread spectrum with RAKE receiver

In Chapter 5, we discussed the main techniques to improve link quality of wireless channels by compensating the fading channel impairments through independent paths, error detection or correction coding or equalization. However, besides compensating the channel fading, other methods can also be taken to improve performance of wireless communications.

One method is spread spectrum, where the performance can be improved by increasing the transmit signal bandwidth. Spread spectrum expands the signal bandwidth beyond the minimum bandwidth necessary for data transmission (Scholtz 1982). The spread spectrum techniques have the following advantages:

(1) The signal is hidden below the noise floor and this makes the signal difficult to detect.
(2) Spread spectrum also mitigates the performance degradation due to inter-symbol interference (ISI) and narrowband interference.
(3) Spread spectrum with a RAKE receiver can provide coherent combining of different multipath components.
(4) Spread spectrum allows multiple users to share the same signal frequency band, since these signals can be superimposed on top of each other and demodulated with minimal interference.
(5) The wide bandwidth of spread spectrum signals is useful for location and timing acquisition.

In this chapter, we will discuss the principle of spread spectrum system. Since direct sequence spread spectrum (DSSS) lays the foundation of code division multiple access (CDMA) because of the correlation property of pseudonoise (PN) code, we spend a lot of time discussing the generation of PN codes. We also discuss how to use RAKE receiver to improve the performance of a wireless system, which can be regarded as one kind of frequency diversity.

6.1 Spread spectrum

6.1.1 Principle of spread spectrum

The fundamental of spread spectrum communication can be explained by the famous Shannon capacity formula

$$C = B\log_2(1+S/N) \tag{6.1}$$

https://doi.org/10.1515/9783110751437-006

where C is the capacity, B is the bandwidth, S is the signal power and N is the noise power within the signal bandwidth. Equation (6.1) shows the capacity can be kept unchanged by increasing the bandwidth of the signal if the signal-to-noise power ratio (SNR) is decreased, which means no matter how low the SNR is, as long as the signal bandwidth is extended sufficiently, the same information transmission rate can be maintained to transmit information reliably. This is the theoretical basis of spread spectrum communication.

Spread spectrum firstly found applications in military area because of its hiding property of the signal below the noise floor during transmission, its resistance to narrowband jamming and interference, and its low probability of detection and interception. It was then used in commercial applications; for example, it was used in cordless phones since its narrowband interference resistance property and in cellular systems and wireless local area networks (LANs) since its ISI rejection and bandwidth sharing capabilities. Spread spectrum laid the basis for third-generation cellular systems as well as second-generation wireless LANs.

DSSS system is the most typical spread spectrum communication system. Figure 6.1 shows the DSSS model.

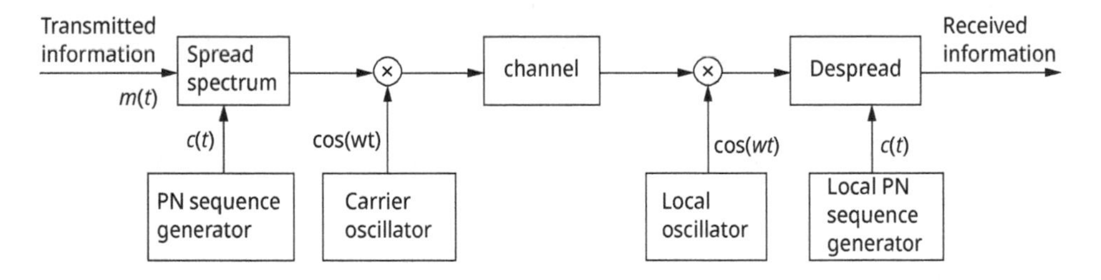

Fig. 6.1: Direct sequence spread spectrum (DSSS) communication system.

From Fig. 6.1, we can see that there is a spread spectrum process at the transmitter and a de-spread process at the receiver in the DSSS communication model. In the transmitter, a PN code with a rate of far above that of the information is used to spread the transmitted information to a very high chip rate, which occupies a very big bandwidth. This high-speed chip will be modulated to a high-frequency carrier. At the receiver, the signal is first de-modulated by a local carrier frequency, and then de-spread by the same PN code; in such a way, the signal energy spread to a high frequency is collected back to the narrow band.

The signals in different steps of the DSSS system are shown in Fig. 6.2. In the receiver, the wide signal correlated with the PN codes will be reduced to the original narrowband signal by de-spreading, whereas the broadband noise which is uncorrelated with PN codes will be still maintained in broadband. The wideband signal after de-spreading is then filtered by a narrowband filter, and the useful signal is obtained,

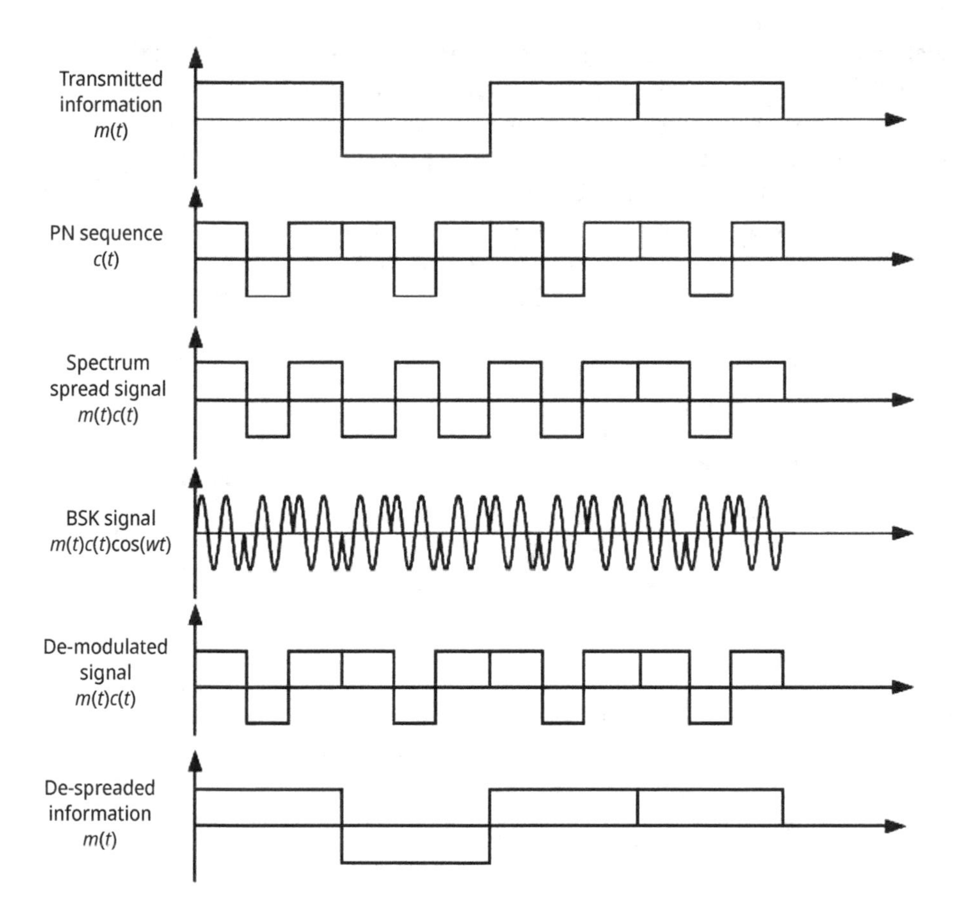

Fig. 6.2: Direct sequence spread spectrum modulation signals.

while most of the noise signal is filtered out, which greatly improves the signal-to-noise ratio and reduces the bit error rate.

6.1.2 Characteristics of spread spectrum communications

In spread spectrum communications, the signal is spread by a PN code at the transmitter and it is correlated by the same PN code at the receiver, which has many advantages:

(1) Strong anti-interference capability
In the transmitter, the signal is spread to a large bandwidth, and then de-spread by the same PN code. This de-correlation compressed the bandwidth and recovered the narrowband signal, and only the signal correlated with the PN code can get a high output peak, whereas other undesired signals, including narrow band interference,

wideband interference and man-made interference, will still remain on the large bandwidth and low power density because they are uncorrelated to the PN code. In this way, the interference power entering the signal bandwidth is greatly reduced, and the useful signal can be separated with most of the undesired signals filtered out. As a result, the signal to interference plus noise power ratio of the output is increased accordingly. Therefore, in the environment of electronic countermeasures, the spread spectrum system has a very strong anti-jamming ability which is unachievable for all other communication systems. Figure 6.3 shows the principle of anti-interference.

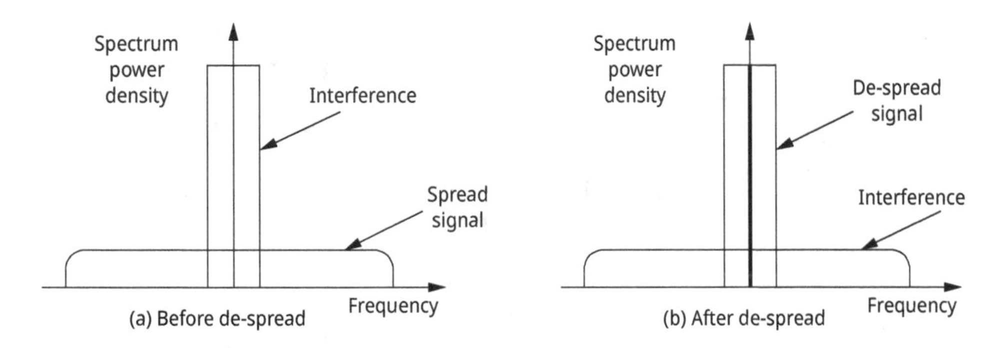

Fig. 6.3: Power spectrum density of spread signal and interference before and after de-spread.

The anti-interference capability of a spread system is determined by its processing gain, which is also called spreading gain, defined by

$$G = \frac{B_{\text{PN}}}{B_{\text{information}}} = \frac{T_{\text{information}}}{T_{\text{PN}}} \tag{6.2}$$

where B and T denote the bandwidth/rate and symbol length. For a given constant information rate, the higher rate of PN sequence will result in a wider spread signal bandwidth, and this yields a greater processing gain and a stronger anti-jamming ability.

(2) High capability of anti-fading
The bandwidth occupied by spread spectrum signals is generally very high. When there is a fading caused by some reason, such as frequency-selective fading, the fading will affect a small part of the bandwidth and will not cause distortion of the whole signal. Therefore, the spread spectrum system has the ability to resist frequency-selective fading.

(3) Strong resistance to multipath interference
Since the pseudo-code used in the spread spectrum system usually has good autocorrelation characteristics, the signals transmitted by different paths can be easily separated, and then the use of diversity reception technology to form superposition of several paths can greatly improve the performance of the system.

(4) Low probability of interception
The signal transmitted by the spread spectrum system is expanded to a wide frequency band with a very low spectral density, which is similar to white noise. Some systems can work under the condition of −20 to −15 dB. It is difficult to find the existence of signal for other systems, and it is even more difficult to extract useful information without knowing the spread spectrum coding. Therefore, it has a low probability of interception. At the same time, the encryption key can be used to encrypt the information to enhance the security and confidentiality of the spread spectrum system.

(5) Easy to implement CDMA
Although spread spectrum communication occupies broadband spectrum resources, its strong multiple access ability ensures its high spectrum utilization. Because of the same spread spectrum code used in both transmitter and receiver, it has the potential for frequency multiplexing and multiple access communications (Kohno et al. 1995). By using the irrelevant spread spectrum codes for different users, the signals of different users can be distinguished. By using their own spread spectrum codes, they can communicate at the same frequency band at the same time without interfering with each other, thus realizing frequency multiplexing. This multi-access mode is flexible, fast and suitable for flexible networking, which is suitable for tactical communications and mobile communications.

From Fig. 6.1, we know that the main difference between spread spectrum system and non-spread spectrum system is the spread and de-spread procedure, and the property of the PN code plays a key role in this procedure, which is also very important in CDMA-based system, so we discuss PN sequence in the following section.

6.1.3 Pseudo-random sequences in spread spectrum systems

6.1.3.1 The concept of pseudo-random codes

According to Shannon's coding theorem, as long as the information rate R_a is less than the channel capacity C, some coding method can always be found, so that the original information can be transmitted over additive white Gaussian noise (AWGN) channel with error approaching to zero under the condition of a fairly long code word.

(1) Shift register sequence
Binary sequence is often used in engineering. The elements in the sequence only have two values of "0" or "1," which correspond to two levels of the electrical signal, positive and negative. Binary sequences are generally generated by shift registers, and the resulting sequence is called a shift register sequence. The structure of the shift register sequence generator is shown in Fig. 6.4, which is called the simple shift register generator (SSRG). The sequence generated by SSRG as shown in Fig. 6.4 is

10 00 00 10 00 01 1010 10 11 11 1010 11 11 11 11 11 11 11 11 11 11 11 1010 1010 11 11 1

The sequence has a total length of 63 bits, that is, its period of 63.

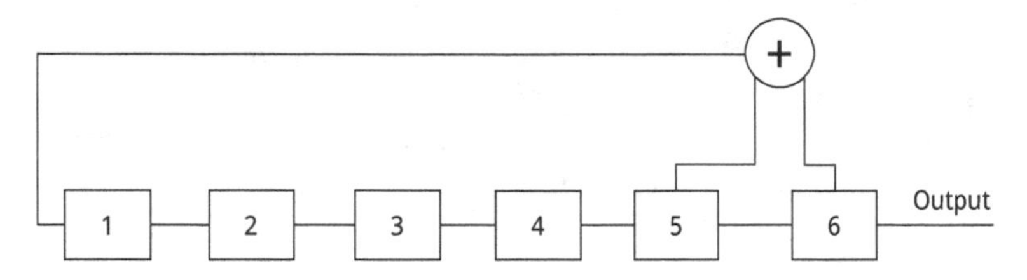

Fig. 6.4: The shift register sequence generator.

Multi-return shift register generator (MSRG) is another type of shift register sequence generator. An example of MSRG is shown in Fig. 6.5.

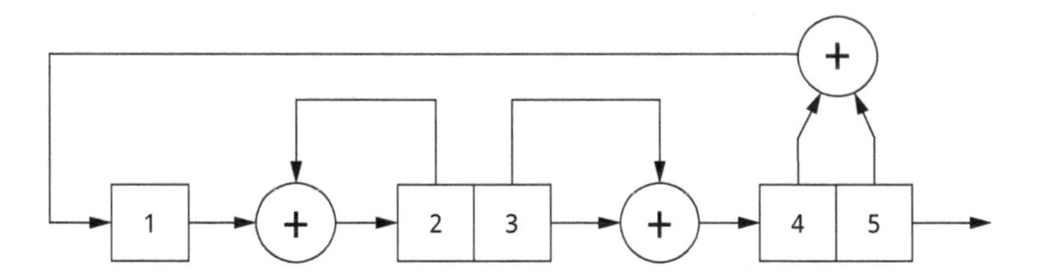

Fig. 6.5: An example of MSRG.

In the generators, the additions used are modulo 2 additions, where the rule is shown in Fig. 6.6.

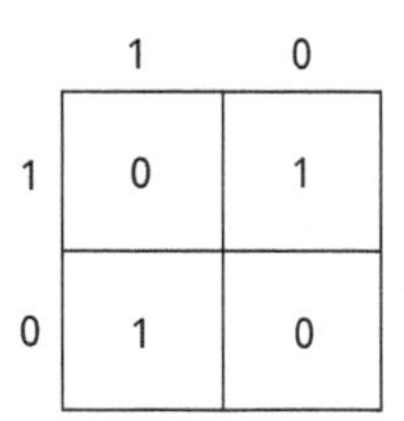

Fig. 6.6: Rule for modulo 2 additions.

(2) Correlation properties of sequences

For the pseudo-random sequence in the spread spectrum system, the most important thing is the correlation properties, including autocorrelation, cross-correlation and characteristic and partial correlation. The definitions of these correlation functions are given below.

Given there are two sequences of $\{a\}$ and $\{b\}$ of length N, the elements in the sequence are a_i and b_i, respectively, $i = 0, 1, 2, 3, 4, \ldots, N-1$; then the autocorrelation function $R_a(j)$ of the sequence is defined as

$$R_a(j) = \sum_{i=0}^{N-1} a_i a_{i+j} \tag{6.3}$$

Since $\{a\}$ is a periodic sequence, $a_{N+i} = a_i$. Its autocorrelation coefficient $\rho_a(j)$ is defined as

$$\rho_a(j) = \frac{1}{N}\sum_{i=0}^{N-1} a_i a_{i+j} \tag{6.4}$$

The cross-correlation function $R_{ab}(j)$ of sequences $\{a\}$ and $\{b\}$ is defined as

$$R_{ab}(j) = \sum_{i=0}^{N-1} a_i b_{i+j} \tag{6.5}$$

The corresponding cross-correlation coefficient is defined as

$$\rho_{ab}(j) = \frac{1}{N}\sum_{i=0}^{N-1} a_i b_{i+j} \tag{6.6}$$

For the binary sequences, it can be expressed as

$$\rho_{ab}(j) = \frac{A-D}{N} \tag{6.7}$$

where A is the number of corresponding elements for $\{a\}$ and $\{b\}$ that are the same, D is the number of corresponding elements for $\{a\}$ and $\{b\}$ that are different.

If $\rho_{ab}(j) = 0$, we define $\{a\}$ is orthogonal to sequence $\{b\}$.

The partial correlation function and the partial correlation coefficient for sequence $\{a\}$ are defined, respectively, as follows:

$$R_{aP}(j) = \sum_{i=t}^{P+t-1} a_i a_{i+j} \qquad P \le N \tag{6.8}$$

$$\rho_{abP}(j) = \frac{1}{N}\sum_{i=t}^{P+t-1} a_i a_{i+j} \qquad P \le N \tag{6.9}$$

where t is a constant.

The partial cross-correlation function and partial cross-correlation coefficient of sequences $\{a\}$ and $\{b\}$ are defined, respectively:

$$R_{abP}(j) = \sum_{i=t}^{P+t-1} a_i a_{i+j} \qquad P \le N \tag{6.10}$$

$$\rho_{abP}(j) = \frac{1}{N}\sum_{i=t}^{P+t-1} a_i a_{i+j} \qquad P \le N \tag{6.11}$$

(3) Definition of PN code

White noise is a random process, and the instantaneous value is normally distributed. Its autocorrelation function and power spectral density as shown in formulas (6.12) and (6.13) are as follows:

$$R_a(\tau) = \frac{n_0}{2}\delta(\tau) \tag{6.12}$$

$$G_n(\omega) = \frac{n_0}{2} \tag{6.13}$$

which shows an excellent autocorrelation property.

Pseudo-random sequences are evolved from the white noise. By using the coding structure, there are only two levels of "0" and "1." Therefore, the probability distribution of PN coding is not exactly normal distribution. However, when the length of the code is big enough, it approaches normal distribution according to the central limit theorem. The pseudo-random code is defined as follows:

(1) Codes with the following form of autocorrelation coefficients are called narrow-sense pseudo-random codes:

$$\rho_a(j) = \begin{cases} \frac{1}{N}\sum\limits_{i=0}^{N-1} a_i^2 = 1 & j = 0 \\ \frac{1}{N}\sum\limits_{i=0}^{N-1} a_i a_{i+j} = -\frac{1}{N} & j \neq 0 \end{cases} \tag{6.14}$$

(2) Codes with autocorrelation coefficients in the following form are called generalized pseudo-random codes of the first kind:

$$\rho_a(j) = \begin{cases} \frac{1}{N}\sum\limits_{i=0}^{N-1} a_i^2 = 1 & j = 0 \\ \frac{1}{N}\sum\limits_{i=0}^{N-1} a_i a_{i+j} = c < 1 & j \neq 0 \end{cases} \tag{6.15}$$

(3) Codes with autocorrelation coefficients in the following form are called generalized pseudo-random codes of the second kind:

$$\rho_{ab}(j) \approx 0 \tag{6.16}$$

(4) The code whose correlation function satisfies one of eqs. (6.14)–(6.16) is called pseudo-random code.

6.1.3.2 Generating of *m* sequence

(1) Feedback shift register

The m sequence is the longest linear shift register sequence, which is formed by the shift register plus feedback. Its structure is shown in Fig. 6.7.

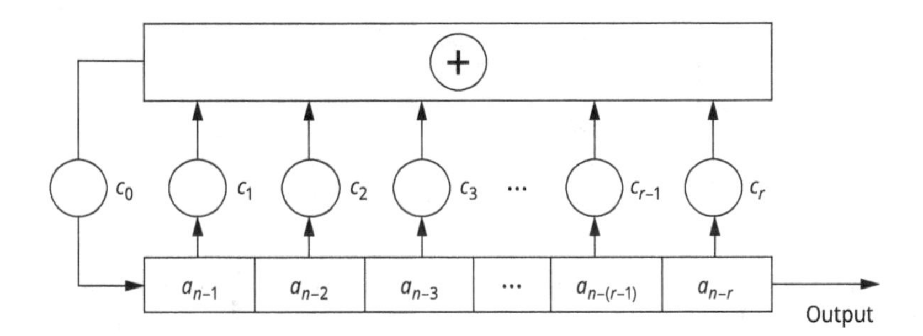

Fig. 6.7: Feedback shift register.

(2) Cyclic sequence generator

The longest linear shift register sequence can be obtained by the recursive relation of feedback logic. In understanding the principle, we need to introduce the concept of sequence polynomial.

(i) Sequence polynomial

A polynomial with the elements a_n ($n = 0, 1, \ldots$) of a bivariate finite field as coefficients is called the generating polynomial of sequence or sequence polynomial:

$$G(x) = a_0 + a_1x + a_2x^2 + \cdots + a_nx^n + \cdots = \sum_{n=0}^{\infty} a_nx^n \tag{6.17}$$

For a feedback shift register, the sequence generated is determined as soon as the feedback logic is determined. So what is the relationship between sequence and feedback logic? As shown in Fig. 6.7, the state of the first shift register at the next moment is determined jointly by the state feedback of the r shift registers at this time, i.e.

$$a_n = c_1a_{n-1} + c_2a_{n-2} + c_3a_{n-3} + \cdots + c_ra_{n-r} = \sum_{i=1}^{r} c_ia_{n-i} \tag{6.18}$$

Thus, the sequence satisfies the linear recursive relation. Move a_n to the right of the equation and consider $c_0 = 1$, we get

$$c_na_n + \sum_{i=1}^{r} c_ia_{n-i} = \sum_{i=0}^{r} c_ia_{n-i} \tag{6.19}$$

(ii) Characteristic polynomial

A feedback shift register can also be described by matrix A, called state transition matrix, which is of order $r \times r$ as follows:

$$A = \begin{bmatrix} c_1 & c_2 & c_3 & \cdots & c_{r-1} & 1 \\ 1 & 0 & 0 & \cdots & 0 & 0 \\ 0 & 1 & 0 & \cdots & 0 & 0 \\ \vdots & \vdots & \vdots & \vdots & & \vdots \\ 0 & 0 & 0 & \cdots & 1 & 0 \end{bmatrix} \tag{6.20}$$

As given in eq. (6.20), the elements in first row of A are the feedback logic of the shift register. Where $c_r = 1$, the sub-matrix except the first row and the r column is a unit matrix of $(r-1) \times (r-1)$. Thus, the structure of the A matrix and shift register corresponds one to one. The A matrix can associate the next state of the shift register with the current state.

Let the current state and the next state of shift register be represented by vectors a_n and a_{n+1}, respectively:

$$a_n = \begin{bmatrix} a_{n-1} \\ a_{n-2} \\ a_{n-3} \\ \vdots \\ a_{n-r} \end{bmatrix} \quad a_{n+1} = \begin{bmatrix} a_{(n+1)-1} \\ a_{(n+1)-2} \\ a_{(n+1)-3} \\ \vdots \\ a_{(n+1)-r} \end{bmatrix} \tag{6.21}$$

Then

$$a_{n+1} = Aa_n \tag{6.22}$$

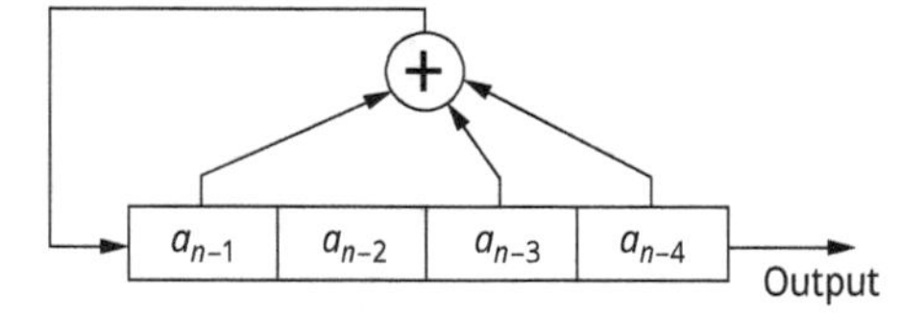

Fig. 6.8: An example of feedback shift register.

For the example given in Fig. 6.8,

$$A = \begin{bmatrix} 1 & 0 & 1 & 1 \\ 1 & 0 & 0 & 0 \\ 0 & 1 & 0 & 0 \\ 0 & 0 & 1 & 0 \end{bmatrix} \tag{6.23}$$

$$\begin{bmatrix} a_{(n+1)-1} \\ a_{(n+1)-2} \\ a_{(n+1)-3} \\ a_{(n+1)-4} \end{bmatrix} = \begin{bmatrix} 1 & 0 & 1 & 1 \\ 1 & 0 & 0 & 0 \\ 0 & 1 & 0 & 0 \\ 0 & 0 & 1 & 0 \end{bmatrix} \begin{bmatrix} a_{n-1} \\ a_{n-2} \\ a_{n-3} \\ a_{n-4} \end{bmatrix} \tag{6.24}$$

That is,

$$\left.\begin{aligned} a_{(n+1)-1} &= a_{n-1} + a_{n-3} + a_{n-4} \\ a_{(n+1)-2} &= a_{n-1} \\ a_{(n+1)-3} &= a_{n-2} \\ a_{(n+1)-4} &= a_{n-3} \end{aligned}\right\} \tag{6.25}$$

(iii) The relation between characteristic polynomial and sequence polynomial
Assume that the linear shift register sequence is

$$\{a_n\} = a_0, a_1, a_2, \ldots, a_n, \ldots$$

The corresponding sequence polynomial is

$$G(x) = \sum_{n=0}^{\infty} a_n x^n \tag{6.26}$$

The linear recursive feedback function of $\{a_n\}$ is shown in eq. (6.18), which is rewritten as follows:

$$a_n = \sum_{n=1}^{r} c_i a_{n-i} \tag{6.27}$$

We get

$$G(x) = \sum_{n=0}^{\infty} \left[\sum_{i=1}^{r} c_i a_{n-i} \right] x^n \tag{6.28}$$

Exchanging the order of summation and performing variable substitution yield

$$
\begin{aligned}
G(x) &= \sum_{n=0}^{\infty} c_i \left[\sum_{i=1}^{r} c_i a_{n-i} x^n \right] \\
&= \sum_{i=1}^{r} c_i x^i \left[\sum_{i=0}^{\infty} a_{n-i} x^{n-i} \right] \\
&= \sum_{i=1}^{r} c_i x^i \left[\sum_{m=-i}^{\infty} a_m x^m \right] \\
&= \sum_{i=1}^{r} c_i x^i \left[\sum_{m=0}^{\infty} a_m x^m + \sum_{m=-i}^{-1} a_m x^m \right] \\
&= \sum_{i=1}^{r} c_i x^i \left[G(x) + \sum_{m=-i}^{-1} a_m x^m \right]
\end{aligned} \tag{6.29}
$$

Considering $c_0 = 1$ and rewriting eq. (6.29) yield

$$
G(x) = \frac{\sum\limits_{i=1}^{r} c_i x^i \left[\sum\limits_{m=-i}^{-1} a_m x^m \right]}{\sum\limits_{i=1}^{r} c_i x^i + 1} = \frac{\sum\limits_{i=1}^{r} c_i x^i \left[\sum\limits_{m=-i}^{-1} a_m x^m \right]}{\sum\limits_{i=0}^{r} c_i x^i} \tag{6.30}
$$

The initial state of the shift register is set to be $a_{-r} = 1$, $a_{-r+1} = \cdots = a_{-2} = a_{-1} = 0$, the numerator of eq. (6.30) is

$$
\sum_{i=1}^{r} c_i x^i \left[\sum_{m=-i}^{-1} a_m x^m \right] = c_r \tag{6.31}
$$

and then,

$$
G(x) = \frac{c_r}{\sum_{i=0}^{r} c_i x^i} = \frac{c_r}{f(x)}
$$

where c_r makes sense only when it is equal to 1. So the relation between sequential polynomial and characteristic polynomial can be obtained:

$$
G(x) = \frac{1}{f(x)} \tag{6.32}
$$

Theorem 6.1: *If the period of the sequence $\{a_n\}$ is N, then f(x) can divide $1 + x^N$, that is, $f(x) | (1 + x^N)$.*

The proof of this theorem is omitted, and interested readers can refer to the relevant literature.

After discussion of these concepts related to characteristic polynomial and sequence polynomial, let's see the conditions for generation of the m sequence.

(3) Conditions for generating m sequence

The following are conditions for generating m sequences:

(i) The period of the code generated by the shift register of order r is $N = 2^r - 1$, and the characteristic polynomial is necessarily irreducible, that is, the longest sequence cannot be factored again. Therefore, feedback tap cannot be arbitrarily decided; otherwise, short code will be generated.

(ii) $1 + x^N$ must be dividable by the irreducible polynomial $f(x)$ of order $r > 1$ because $a^N(x) = (1 + x^N)/f(x)$.

Tab. 6.1: The length of m sequence, the number of irreducible polynomials and the number of m sequences.

r	2^r–1	N_m	N_1
1	1	1	2
2	3	1	1
3	7	2	2
4	15	2	3
5	31	6	6
6	63	6	9
7	127	18	18
8	255	16	30
9	511	48	56
10	1023	60	99
11	2047	176	186
12	4095	144	335
13	8191	630	630
14	16,383	756	1161
15	32,767	1800	2182
16	65,535	2048	4080
17	131,071	7710	7710
18	262,143	8064	14,532
19	524,287	27,594	27,594
20	1,048,575	24,000	52,377
21	2,098,151	84,672	99,858
22	4,194,303	120,032	190,557
23	8,388,607	356,960	364,722
24	16,777,215	276,480	698,870

(iii) If $2^r - 1$ is a prime number, then the linear shift register sequence generated by all irreducible polynomials of order r must be an m sequence, and the irreducible polynomials that produce this m sequence are called primitive polynomials.

(iv) Besides order r, if there is a feedback structure with even taps, the generated sequence is not the longest linear shift register sequence.

(4) The number of irreducible polynomials N_I and the number of m sequences N_m
From the analysis above, we can see that when $2^r - 1$ is prime, all irreducible polynomials of order r decomposed from $1 + x^N$ are characteristic polynomials of m sequence. In this section, we will give the number N_I of irreducible polynomials of order r decomposed from $1 + x^N$ and the number N_m of characteristic polynomials that can generate m sequences.

According to the unique decomposition theorem, any positive integer n greater than 1 can be expressed as the product of prime numbers, that is

$$n = \prod_{i=1}^{k} p_i^{a_i} \tag{6.33}$$

where p_i and $a_i i = 1, 2, \ldots, k$ are prime numbers and positive power, respectively. The length of m sequence, the number of irreducible polynomials and the number of m sequences are listed in Tab. 6.1.

(5) Feedback coefficient of m sequence
Whether a linear feedback shift register can generate m sequence depends on its circuit feedback coefficient c_i, which is its recursive relation. Different feedback coefficients produce different shift register sequences. Table 6.2 lists the feedback coefficients of the longest linear shift register sequences of different series. For r greater than 9, since there are a large number of m sequences, it is impossible to list them all here, and only a part is listed in the table. For more details, please refer to the relative references.

Tab. 6.2: Feedback coefficients list of m sequence.

Order *r*	Length *N*	Feedback coefficients
3	7	13
4	15	23
5	31	45, 67, 75
6	63	103, 147, 155
7	127	203, 211, 217, 235, 277, 313, 325, 324, 367
8	255	435, 453, 537, 543, 545, 551, 703, 747
9	511	1021, 2033, 2157, 2443, 2745, 3471
10	1023	2011, 2033, 2157, 2443, 2745, 3471
11	2047	4005, 4445, 5023, 5263, 6211, 7363
12	4095	10,123, 11,417, 12,515, 13,505, 14,127, 15,053
13	8191	20,033, 23,261, 24,633, 30,741, 32,535, 37,505
14	16,383	42,103, 51,761, 55,753, 60,153, 71,147, 67,401
15	32,767	100,003, 110,013, 120,265, 133,663, 142,305, 164,705
16	65,535	210,013, 233,303, 307,572, 311,405, 347,433, 375,213
17	131,071	400,011, 411,335, 444,257, 527,427, 646,775, 714,303
18	262,143	10,000,201, 1,002,241, 1,025,711, 1,703,601
19	524,287	2,000,047, 2,020,471, 2,227,023, 2,331,067, 2,570,103, 3,610,353
20	1,048,575	4,000,011, 4,001,151, 4,004,515, 442,235, 6,000,031

The number of feedback coefficients in Tab. 6.2 is octal. After converting these numbers into binary forms, the corresponding feedback coefficients can be obtained. For example, if $r = 9$, the feedback coefficient is 1157, which can be converted to the binary number and it corresponds to shift register, that is:

$$\begin{matrix} c_9 & c_8 & c_7 & c_6 & c_5 & c_4 & c_3 & c_2 & c_1 & c_0 \\ 1 & 0 & 0 & 1 & 1 & 0 & 1 & 1 & 1 & 1 \end{matrix}$$

$c_9 = c_6 = c_5 = c_3 = c_2 = c_1 = c_0 = 1$ means there are feedbacks at the corresponding registers, and $c_8 = c_7 = c_4 = 0$ means no feedback. At the same time, the feedback coefficients of the characteristic polynomials for generating m sequences relative to 1157 can be obtained. The characteristic polynomial is

$$f(x) = x^9 + x^6 + x^5 + x^3 + x^2 + x + 1 \tag{6.34}$$

Through the feedback coefficients in Tab. 6.2, the feedback taps and characteristic polynomials of the corresponding mirror sequence can also be obtained. The so-called mirror sequence is the sequence opposite to the original sequence. If the sequence of $R = 3$ is 1110100, the sequence of mirrors is 0010111. The characteristic polynomial $f^{(R)}(x)$ of the mirror image sequence can be obtained from the characteristic polynomial $f(x)$ of the original sequence, i.e.

$$f^{(R)}(x) = x^r f\left(\frac{1}{x}\right) \tag{6.35}$$

Figure 6.9 gives a structure of the original sequence and the corresponding mirror sequence of $r = 7$.

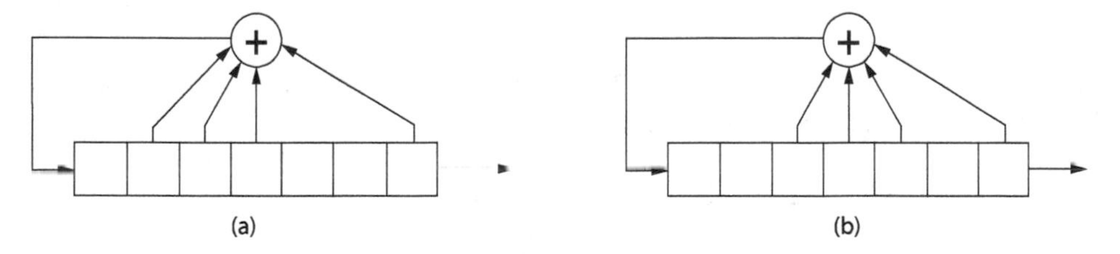

Fig. 6.9: The structure of the original sequence and mirror sequence of $r = 7$: (a) original sequence and (b) mirror sequence.

(6) The structure of m sequence generator
The structure of m sequence generator generally has two forms, simple type (SSRG) and module tap type (MSRG), which will be descried below:

(i) SSRG

The structure of SSRG is shown in Fig. 6.10. The feedback logic of this structure is determined by the characteristic polynomial. The disadvantage of this structure is that the device delay in the feedback branch is superimposed, which is equal to the sum of all modulus 2 adder delays in the feedback branch. Therefore, the speed of pseudo-random sequence is limited. One way to improve SSRG's working speed is to select m sequences with fewer taps, which can also simplify the structure of sequence generators.

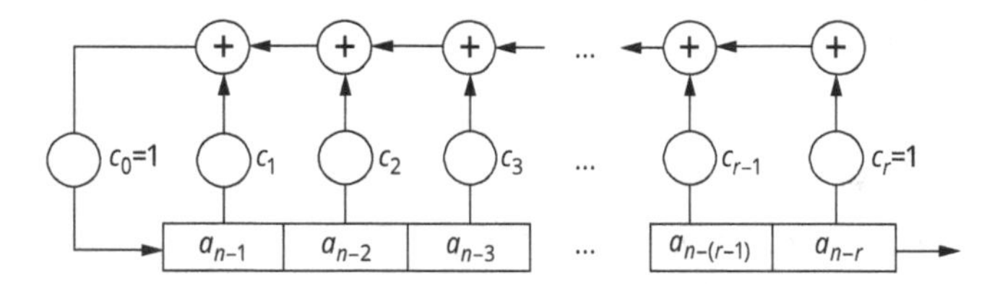

Fig. 6.10: SSRG structure.

(ii) MSRG

Another way to improve the working speed of pseudo-random sequence is to adopt the MSRG structure. The structure of this sequence generator is shown in Fig. 6.11. The characteristic of this structure is that there is a module 2 adder between each level trigger and its adjacent level trigger, and there is no delay component in the feedback path. This type of sequence generator has been modularized. The total feedback delay of this structure is only the delay time of an adder of modulo 2, so it can improve the working speed of the generator. The highest operating frequency of SSRG sequence generator is

$$f_{\max} = \frac{1}{T_R + \sum T_M} \tag{6.36}$$

where T_R is the transmission delay of the first-order shift register and $\sum T_{\mathrm{M}}$ is the sum of mode 2 plus delay in the feedback network.

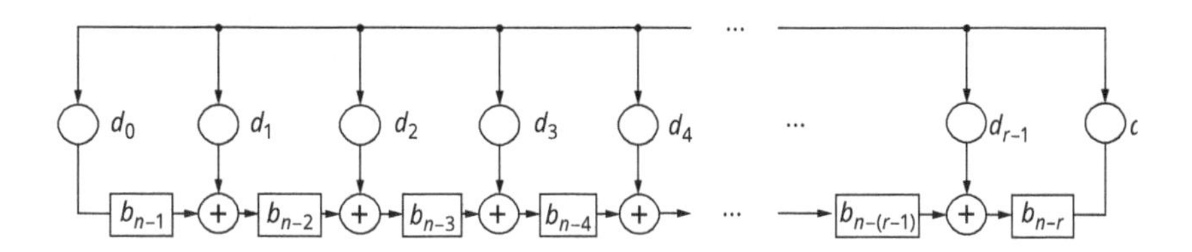

Fig. 6.11: MSRG structure.

The maximum operating frequency of MSRG sequence generator is

$$f_{\max} = \frac{1}{T_R + T_M} \tag{6.37}$$

where T_M is the transmission delay of the first-order modulo 2 adders.

Figure 6.12 gives two kinds of sequence generators when $r = 6$. Table 6.3 gives the corresponding state transition of the two structures.

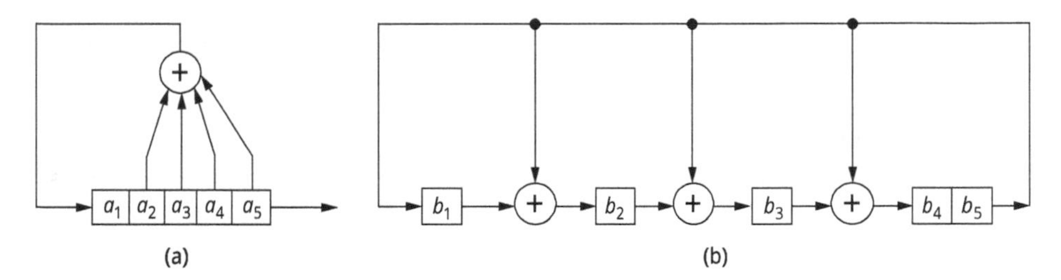

Fig. 6.12: Sequence generators with $r = 5$: (a) SSRG structure and (b) MSRG structure.

Tab. 6.3: State transition tables with SSRG and MSRG structures.

	SSRG	MSRG		SSRG	MSRG
	$a_1a_2a_3a_4a_5$ output	b_1 b_2 b_3 b_4 b_5 output		a_1 a_2 a_3 a_4 a_5 output	b_1 b_2 b_3 b_4 b_5 output
1	1 1 1 1 1 1	1 1 1 1 1 1	17	0 1 1 0 1 1	0 1 0 0 0 0
2	0 1 1 1 1 1	1 0 0 0 1 1	18	1 0 1 1 0 0	0 0 1 0 0 0
3	0 0 1 1 1 1	1 0 1 1 0 0	19	0 1 0 1 1 1	0 0 0 1 0 0
4	1 0 0 1 1 1	0 1 0 1 1 1	20	1 0 1 0 1 1	0 0 0 0 1 1
5	0 1 0 0 1 1	1 1 0 1 1 1	21	0 1 0 1 0 0	1 1 1 1 0 0
6	0 0 1 0 0 0	1 0 0 1 1 1	22	0 0 1 0 1 1	0 1 1 1 1 1
7	1 0 0 1 0 0	1 0 1 1 1 1	23	0 0 0 1 0 0	1 1 0 0 1 1
8	1 1 0 0 1 1	1 0 1 0 1 1	24	1 0 0 0 1 1	1 0 0 1 0 0
9	0 1 1 0 0 0	1 0 1 0 0 0	25	1 1 0 0 0 0	0 1 0 0 1 1
10	0 0 1 1 0 0	0 1 0 1 0 0	26	1 1 1 0 0 0	1 1 0 1 0 0
11	0 0 0 1 1 1	0 0 1 0 1 1 1	27	0 1 1 1 0 0	0 1 1 0 1 1
12	0 0 0 0 1 1	1 1 1 0 0 0	28	1 0 1 1 1 1	1 1 0 0 0 0
13	1 0 0 0 0 0	0 1 1 1 0 0	29	1 1 0 1 1 1	0 1 1 0 0 0
14	0 1 0 0 0 0	0 0 1 1 1 1	30	1 1 1 0 1 1	0 0 1 1 0 0
15	1 0 1 0 0 0	1 1 1 0 1 1	31	1 1 1 1 0 0	0 0 0 1 1 1
16	1 1 0 1 0 0	1 0 0 0 0 0	32	1 1 1 1 1 1	1 1 1 1 1 1

6.1.3.3 Properties of *m* sequences

The properties of *m* sequences are very important which can be used in applications.

(1) Main properties of *m* sequences

(i) The balance property

In a period of *m* sequence, the number of "1" and "0" are basically equal. To be precise, the number of "1" is one more than that of "0."

(ii) Run length distribution

We call successive elements with the same values in a sequence a run. In a run, the number of elements is called run length.

(iii) Shift additivity

A sequence $\{a_n\}$ is mod 2 added with another different sequence $\{a_{n+m}\}$ generated by *m* time delay shift, and the result sequence $\{a_{n+k}\}$ is still a delayed shift sequence of $\{a_n\}$, i.e.

$$\{a_n\} + \{a_{n+m}\} = \{a_{n+k}\} \tag{6.38}$$

(iv) Periodicity

The period of *m* sequence is $N = 2^r - 1$, where *r* is the series of feedback shift registers.

(v) Pseudo-randomness

If a normal distribution white noise is sampled and the sampling value is positive, it is marked as "+". If the sampling value is negative, it is marked as "–", then the polarity obtained from each sampling is arranged into a sequence, which can be written as " –"

$$\cdots ++-+--+----+-+--+++-- \cdots$$

This sequence is a random one with the following basic properties:

- The occurrence probability of " +" and "–" in the sequence is equal.
- Runs with length of 1 account for about 1/2 of the sequence, runs with length of 2 account for about 1/4 and runs with length of 3 account for about 1/8.
- Since the white noise power spectrum is constant, the autocorrelation function is a Dirac delta function $\delta(\tau)$.

(2) Correlation properties of *m* sequences

The autocorrelation function of periodic function $s(t)$ is defined as

$$R_s(\tau) = \frac{1}{T}\int_{-T/2}^{T/2} s(t)s(t+\tau)d\tau \tag{6.39}$$

where *T* is the period of $s(t)$.

For binary code sequence $\{a_n\}$ with "1" and "0," the value of autocorrelation function is

$$R(j) = \sum_{i=0}^{N-1} a_i a_{i+j} \tag{6.40}$$

The correlation coefficient is

$$\rho(j) = \frac{1}{N}\sum_{i=0}^{N-1} a_i a_{i+j} = \frac{A-D}{N} \tag{6.41}$$

Figure 6.13 shows the waveform of $R(\tau)$. When the period NT_c is very long and the code width T_c is very small, $R(\tau)$ is similar to the shape of the impulse function $\delta(\tau)$.

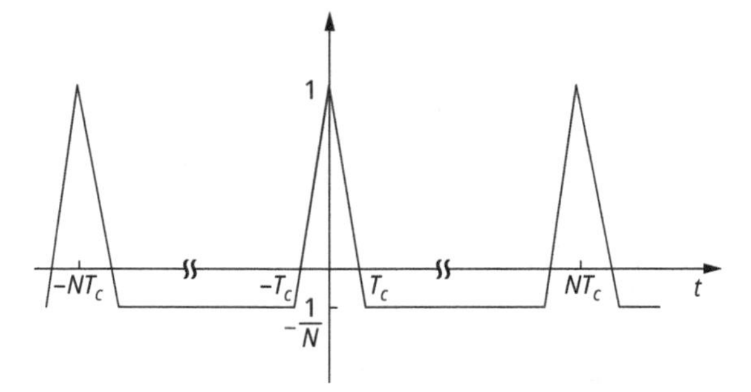

Fig. 6.13: Autocorrelation functions of *m* sequences.

(3) Power spectrum of *m* sequence

As we know, a Fourier transform pair is formed between the autocorrelation function and the power spectrum of the signal:

$$\begin{cases} G(\omega) = \int_{-\infty}^{\infty} R(\tau)e^{-j\omega\tau}d\tau \\ R(\tau) = \frac{1}{2\pi}\int_{-\infty}^{\infty} G(\omega)e^{j\omega\tau}d\omega \end{cases} \tag{6.42}$$

Since the autocorrelation function of *m* sequence is periodic, the corresponding spectrum is discrete. The waveform of the autocorrelation function is triangular wave, and the envelope of the corresponding discrete spectrum is $Sa^2(x)$. The power spectrum $G(\omega)$ of the *m* sequence can be obtained:

$$G(\omega) = \frac{1}{N^2}\delta(\omega) + \frac{N+1}{N^2}Sa^2\left(\frac{\omega T_c}{2}\right)\sum_{\substack{k=-\infty \\ k\neq 0}}^{\infty}\delta\left(\omega - \frac{2k\pi}{NT_c}\right) \tag{6.43}$$

Figure 6.14 shows the spectrum of $G(\omega)$, and T_c is the duration of PN code chip.

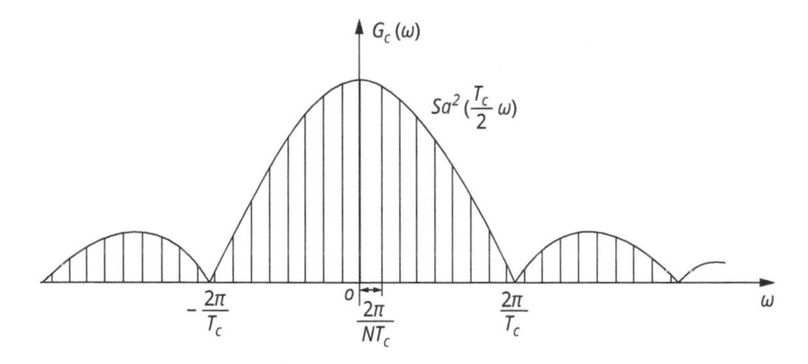

Fig. 6.14: Spectrum of *m* sequence.

From eq. (6.43) and Fig. 6.14, we get to know the following results:
(1) The power spectrum of the *m* sequence is discrete spectrum, and the spectral line spacing is $\omega_1 = 2\pi/(NT_c)$.
(2) The envelope of the power spectrum is $Sa^2(T_c\omega/2N)$, and the power of each component is inversely proportional to the cycle N.
(3) The DC component is inversely proportional to N^2. The larger the N is, the smaller the DC component and the smaller the load leakage would be.
(4) The bandwidth is determined by the symbol width T_c. The smaller the T_c is, the higher the symbol rate and the wider the bandwidth would be.
(5) The first zero point appears at $2\pi/T_c$.
(6) Increasing the length of *m* sequence N and decreasing the symbol width T_c will make the spectral line encrypted and the spectral density reduced, which is closer to the ideal noise characteristics.

6.1.3.4 Gold sequence

Although the *m* sequences have good autocorrelation property, there are not many *m* sequences of the same period, and not all the *m* sequences have good cross-correlation.

In 1967, R. GOLD proposed a code sequence based on the optimal pair of *m* sequence, which was called Gold sequence. Gold sequence has good correlation characteristics and simple generating method; besides, the number of this kind sequence is huge, that's why it finds wide applications in practice. We need to also discuss Gold sequence.

(1) Selection of address code

The purpose of CDMA is to distinguish signal waveforms. Spread spectrum communication is a multi-access communication method that distinguishes communication addresses by code differences (Viterbi 1995, Ziemer et al. 1995). Therefore, the performance of access code is directly related to the performance of the system. Generally speaking, the multiple access codes of different networks are different. The cross-correlation of accessing codes of different networks should be zero, that is, the accessing codes are orthogonal, i.e.

$$\int_T c_i(t)c_j(t)\mathrm{d}t = \begin{cases} 1 & i=j \\ 0 & i\neq j \end{cases} \tag{6.44}$$

where $c_i(t)$ and $c_j(t)$ are the waveforms of the address code. Equation (6.44) shows that orthogonal codes are codes with very small cross-correlation values for different codes. This kind of code with the above property is the second kind of generalized pseudo-random code.

Generally, requirements for the address codes are as follows:

(i) They have good autocorrelation, cross-correlation and partial correlation properties.
(ii) The number of code sequences is as large.
(iii) There is a certain length.
(iv) It is easy to implement the system synchronization, and the acquisition time is short.
(v) It is easy to implement. The equipment is simple and the cost is low.

Gold sequence is one kind of sequence that is suitable for addressing.

(2) Generation of Gold code
(i) *m* sequence optimum pair
The *m* sequence optimum pair refers to the two *m* sequences whose absolute value of the maximum value of their cross-correlation function $|R_{ab}|_{\max}$ is less than a certain value in the set of *m* sequences.

Let the sequence $\{a\}$ be the *m* sequence generated by the primitive polynomial $f(x)$ of order r, and the sequence $\{b\}$ is the *m* sequence corresponding to the primitive polynomial $g(x)$ of order r. When their cross-correlation function value $R_{ab}(\tau)$ satisfies the inequality

$$|R_{ab}(\tau)| < \begin{cases} 2^{\frac{r+1}{2}}+1, & r \text{ is an odd number} \\ 2^{\frac{r+2}{2}}+1, & r \text{ is even, but it's not divisible by 4} \end{cases} \tag{6.45}$$

Then the *m* sequences $\{a\}$ and $\{b\}$ generated by $f(x)$ and $g(x)$ constitute an optimum pair.

Table 6.4 lists the maximum cross-correlation values of *m* sequence optimum pairs with different code lengths and Tab. 6.5 lists part of the optimization pairs.

(ii) Generation method of Gold code
Gold code is a combination code of *m* sequences. It is obtained by modulo 2 addition of two optimum pairs of m sequences with the same length, the same rate and the different codewords. Gold code sequence owes good autocorrelation and cross-correlation features, and the number of address codes is much larger than that of *m* sequence. An

Tab. 6.4: Maximum cross-correlation values of *m* sequence optimum pairs with different code lengths.

The order of the shift register	Code length	Cross-correlation function values	The normalized value
3	7	≤5	≤5/7
5	31	≤9	≤9/31
6	63	≤17	≤17/63
7	127	≤17	≤17/127
9	511	≤33	≤33/511
10	1023	≤65	≤65/1023
11	2047	≤65	≤65/2047

Tab. 6.5: Part of the optimum pairs.

Order of shift register	Reference primitive polynomial	The corresponding paired primitive polynomial
7	211	217, 235, 277, 325, 203, 357, 301, 323
	217	211, 235, 277, 325, 213, 271, 357, 323
	235	211, 217, 277, 325, 313, 221, 361, 357
	236	277, 203, 313, 345, 221, 361, 271, 375
9	1021	1131, 1333
	1131	1021, 1055, 1225, 1725
	1461	1743, 1541, 1853
10	2415	2011, 3515, 3177
	2641	2517, 2218, 3045
11	4445	4005, 5205, 5337, 5263
	4215	4577, 5747, 6765, 4563

optimum *m* sequence pair can generate $2r+1$ Gold codes. This code generator has a simple structure and is easy to implement, so it is widely used in practice.

Let sequences $\{a\}$ and $\{b\}$ be *m* sequence optimum pairs with length $N=2^r-1$. Taking $\{a\}$ sequence as a reference sequence, the $\{b\}$ sequence is shifted by i times, and the shifted sequence $\{b_i\}$ $(i=0,1,\ldots)$ of $\{b\}$ is obtained. By modulo 2 addition of sequences $\{a\}$ and $\{b_i\}$, a new sequence $\{c_i\}$ of length N is obtained. This new sequence is called Gold sequence, that is,

$$\{c_i\}=\{a\}+\{b_i\},\quad i=0,1,\ldots,N \tag{6.46}$$

Figure 6.15 gives two structures for generating Gold sequence.

(3) The correlation of Gold codes

The 2^r-1 Gold sequences generated by modulo 2 addition of optimum *m* sequence pairs are no longer *m* sequences, and the run-length property and binary correlation

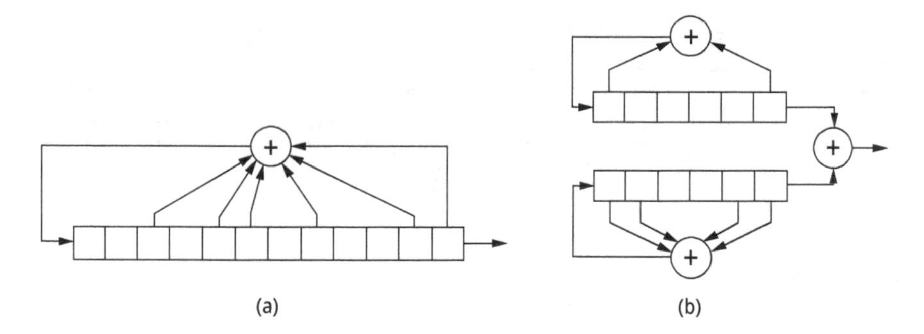

Fig. 6.15: Gold sequence generator: (a) serial structure and (b) parallel structure.

characteristics of *m* sequence will not hold. However, the cross-correlation function between any two Gold sequences satisfies the inequality equation (6.45). Table 6.6 gives the cross-correlations.

(4) Balanced Gold codes

Balanced Gold codes are codes in which the number of "1" is one more than the number of "0." Balanced codes have excellent autocorrelation characteristics. Table 6.7 lists the number of balanced and unbalanced codes with register order of odd number *r*. It can be seen from the table that the number of "1" in the first kind of code sequence is $2^r - 1$, while the number of "0" is $2^r - 2$. The number of "1" is one more than that of "0," so it is a balanced code. The number of this kind of balanced code is $2^{r-1} + 1$. The second and third kinds are unbalanced codes. Let us take $r = 3$ for an example; for the first kind balanced codes, the number of "1" is 4, and there are totally five such sequences. For the second kind, the number of "1" is 6, and there is only one such sequence. For the third kind, the number of "1" is 2, and the number of such sequences is 3.

Balanced codes can suppress the carrier well. Table 6.8 shows the relationship between the balance of codes and carrier suppression.

(5) A method of generating balanced Gold codes

(i) Characteristic phase

In order to find a balanced Gold code, the characteristic phase is first determined. Each longest linear shift register sequence has a characteristic phase. When the sequence is in the characteristic phase, the sequence sampled every other bit is exactly the same as the original sequence, which is the feature of the sequence in the characteristic phase.

Let the characteristic polynomial $f(x)$ of the sequence be a primitive polynomial of the *m* sequence generated by a linear shift register of order *r*. The characteristic phase of the sequence is determined by the ratio of $g(x)/f(x)$, where $g(x)$ is a generating function which is a polynomial with an order equal to or less than *r*. The calculation method of $g(x)$ is as follows:

Tab. 6.6: Cross-correlations of Gold sequences.

Register length	Code length	Normalized value of cross-correlations	Probability of occurrence
r is odd	$N = 2^r - 1$	$-\frac{1}{N}$	0.5
		$-\frac{2^{\frac{r+1}{2}}+1}{N}$	0.25
		$\frac{2^{\frac{r+1}{2}}-1}{N}$	0.25
r is even, but it's not divisible by 4	$N = 2^r - 1$	$-\frac{1}{N}$	0.75
		$-\frac{2^{\frac{r+2}{2}}+1}{N}$	0.125
		$\frac{2^{\frac{r+2}{2}}-1}{N}$	0.125

Tab. 6.7: The number of balanced and unbalanced Gold sequences.

Type	Number of "1" in a sequence	The number of such sequences in the code family
1	2^{r-1}	$2^{r-1}+1$
2	$2^{r-1}+2^{\frac{r-1}{2}}$	$2^{r-2}-2^{\frac{r-3}{2}}$
3	$2^{r-1}-2^{\frac{r-1}{2}}$	$2^{r-2}-2^{\frac{r-3}{2}}$

Tab. 6.8: Relationship between the balance of codes and carrier suppression.

Register order	Code length	The number difference of "1" and "0" in code		Carrier suppression/(dB)	
		Balanced codes	Unbalanced codes	Balanced codes	Unbalanced codes
3	7	1	5	8.45	1.46
5	31	1	9	14.9	5.37
7	127	1	17	21.04	8.37
9	511	1	33	27.08	11.9
11	2047	1	65	33.11	15
13	8191	1	129	39.13	18.03
15	32,767	1	257	45.15	21.06
17	131,071	1	513	51.18	24.07

$$g(x) = \begin{cases} \frac{d[xf(x)]}{dx}, & r \text{ is odd} \\ f(x) + \frac{d[xf(x)]}{dx}, & r \text{ is even} \end{cases} \tag{6.47}$$

The sequence polynomial is

$$G(x) = \frac{g(x)}{f(x)} \tag{6.48}$$

By polynomial quotient, the m sequence in the characteristic phase can be obtained.

(ii) Relative phase
Now let's discuss the relative phase of the shift sequence that produces balanced Gold codes by m sequence optimum pair. Let sequences $\{a\}$ and $\{b\}$ be the m sequence optimum pair in the characteristic phase. When r is odd, its sequence generating polynomial can be expressed as

$$G(x) = \frac{1 + c(x)}{1 + d(x)} \tag{6.49}$$

We can summarize the general steps of generating balanced Gold codes as follows:

Step 1. Choose a reference sequence whose primitive polynomial is $f_a(x)$, and find the generating polynomial $g_a(x)$.

Step 2. Sequence polynomials are obtained from $G(x) = g_a(x)/f_a(x)$, and let sequence $\{a\}$ be in the characteristic phase.

Step 3. Calculate the displacement sequence $\{b\}$ so that the first position of the initial state of the displacement sequence is '0', that is, it is located in the relative phase, corresponding to the first '1' of $\{a\}$.

Step 4. Balanced Gold code sequence can be obtained by modulo 2 addition of sequence $\{a\}$ in the characteristic phase and sequence $\{b\}$ in the relative phase.

6.1.3.5 *M* sequence

Although m sequences are easy to generate, have good autocorrelation characteristics and are pseudo-random, they are small in number and have unsatisfactory cross-correlation characteristics. Because it adopts linear feedback logic, it is easy to decode the sequences of codes, that is, the confidentiality and anti-interception performance are poor.

M sequences are nonlinear sequences, the number of available sequences is large and have good autocorrelation and cross-correlation characteristics, so they are of great significance to discuss m sequence.

There are many methods to construct M sequences, which can be obtained by adding the whole "0" state on the basis of m sequences. It can also be obtained by searching. Whatever method is adopted, it is enough to satisfy that all the 2^r states of the shift register of order r need to be experienced once, and only once, while satisfying the shift register relationship.

(1) M sequence constituted by m sequence

The m sequence contains $2^r - 1$ non-zero states, except for a full "0" state consisting of r "0." Therefore, when M sequence is composed of m sequence, as long as a zero state (r "0") is inserted in an appropriate position, the m sequence with code length of $2^r - 1$ can be increased to the M sequence with code length of 2^r. Obviously, the all-zero state insertion should be after the state of 100 . . . 0, in order to make the all-zero state appear at that time. At the same time, the successor state of the all-zero state must be made to be 00 . . . 01. That is, the state transition process is

$$(000\ldots01) \rightarrow (000\ldots00) \rightarrow (100\ldots00) \tag{6.50}$$

Figure 6.16 gives a four-stage M sequence generator. The corresponding state transition is given in Tab. 6.9.

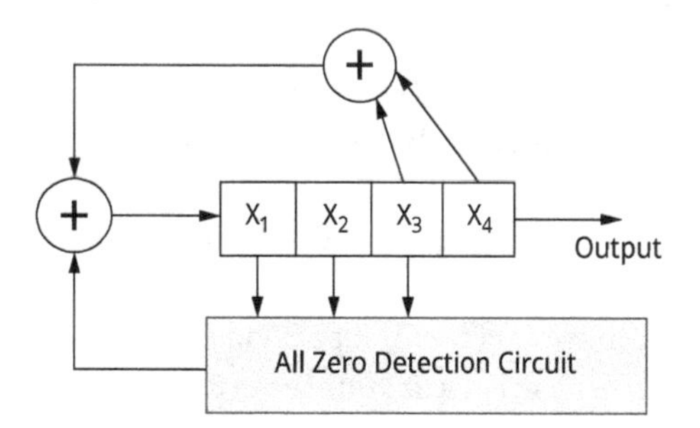

Fig. 6.16: Four-stage *M* sequence generator.

Tab. 6.9: Four-stage *M* sequence state transition.

Timing sequence	$x_1\ x_2\ x_3\ x_4$	Timing sequence	$x_1\ x_2\ x_3\ x_4$
0	1 1 1 1	9	1 1 0 0
1	0 1 1 1	10	0 1 1 0
2	0 0 1 1	11	1 0 1 1
3	0 0 0 1	12	0 1 0 1
4	0 0 0 0	13	1 0 1 0
5	1 0 0 0	14	1 1 0 1
6	0 1 0 0	15	1 1 1 0
7	0 0 1 0	16	1 1 1 1
8	1 0 0 1		

(2) *M* sequence constituted by searching

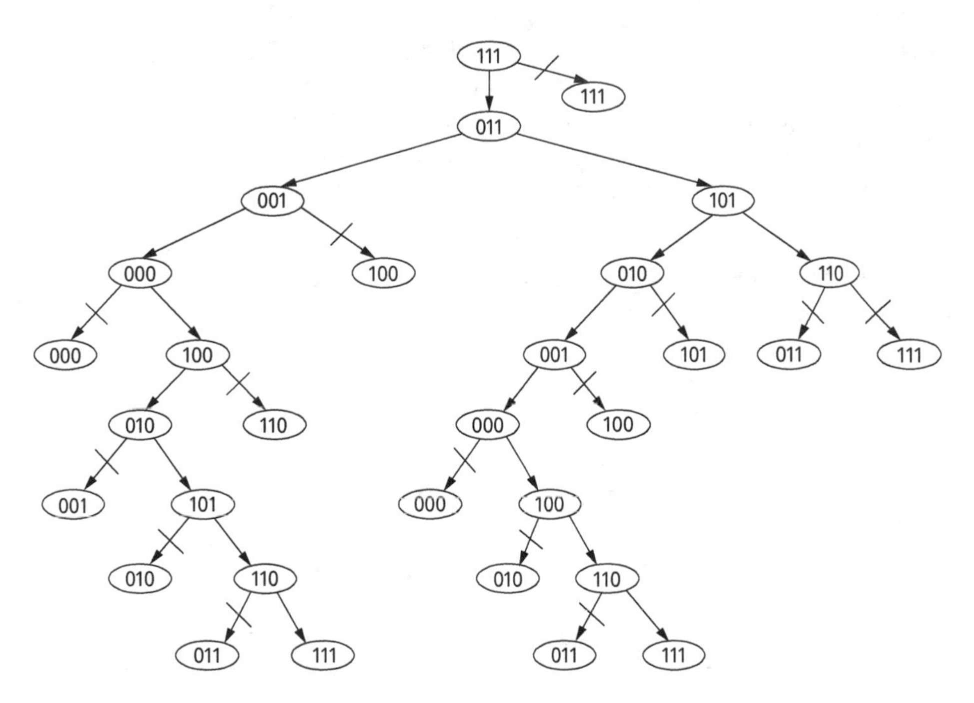

Fig. 6.17: *M* sequence state transition diagram.

The length of *M* sequence is 2^r. It experiences all the 2^r states of the *r*th-order shift register, and each state can only be experienced once. Considering the shifting function of the shift register, the state can be transferred from one state of the *r*th-order shift register, and there is no duplication of states in the transition process. After 2^r transfers, it returns to the state of departure, and a closed loop called Hamilton loop can be obtained. The number of states of the ring is 2^r, from which an *M* sequence can be obtained. Different paths can lead to different *M* sequences. In the case of $r = 3$, the state transition process is shown in Fig. 6.17.

This method can generate *M* sequences generated by all *r*th-order shift registers. As shown in the diagram, only two paths make up a closed loop of $2^r = 8$, that is,

$$(111) \to (011) \to (001) \to (000) \to (100) \to (010) \to (101) \to (110) \to (111)$$

and

$$(111) \to (011) \to (101) \to (010) \to (001) \to (000) \to (100) \to (110) \to (111)$$

The corresponding *M* sequences are 11100010 and 11101000.

(3) Properties of M sequences

(i) The random characteristics of M sequence

There are three main random characteristics of M sequence:

Firstly, the period of M sequence is 2^r, where r is the order of shift register. Secondly, in M sequence with length $N=2^r$, the number of "0" and "1" is the same, i.e., the number of each is 2^{r-1}. Thirdly, the total number of run length is 2^{r-1} in the M sequence of length 2^r, and the numbers of run lengths of "0" and "1" are the same.

(ii) The number of M sequences

The number of M sequences is much more than that of m sequences. The number of M sequences (excluding equivalent translating sequences) is

$$N_M = 2^{2^{r-1}-r} \tag{6.51}$$

Tab. 6.10: The number of m sequences and M sequences.

Order / Sequence type	2	3	4	5	6	7	8	9
m sequences	1	2	2	6	6	18	16	48
M sequences	1	2	16	2048	2^{26}	2^{57}	2^{121}	2^{248}

From Tab. 6.10, we can see that the number of M sequences is far above that of m sequences when $r \geq 4$ so that M sequences as address code can satisfy the requirement of CDMA.

(iii) The correlation properties of M sequences

For any given r-order M sequence, its autocorrelation function $R(\tau)$ is

$$R(\tau) = \begin{cases} 2^r, & \mathrm{r}=0 \\ 0, 1 \leq |\tau| \leq r-1 \\ 2^r - 4\omega(f_0), & \tau \geq r \end{cases} \tag{6.52}$$

where $\omega(f_0)$ is the weight of f_0 and f_0 is the f_0 in $f(x_1, x_2, \ldots, x_r) = x_1 + f_0\,(x_2, x_3, \ldots, x_r)$ which is the feedback function that generates M sequence.

6.2 Spread spectrum with RAKE receiver

In the spread spectrum system, the channel bandwidth is much larger than the channel coherent bandwidth. It is different from the traditional modulation techniques that use equalization we discussed in Chapter 6 to eliminate the ISI between adjacent symbols. In the last section, we have already discussed the autocorrelation characteristics of the spread spectrum code. When spread spectrum is used for CDMA, the good autocorrelation characteristics of spread spectrum code are especially important. In this way, the delay spread in the wireless channel can be regarded as just the re-transmission of the transmitted signal. If the delay between these multipath signals exceeds the length of a chip, they will be regarded as uncorrelated noise by the CDMA receiver, and no longer need to be equalized.

Because of the available information in the multipath signal, the SNR of the received signal can be improved by combining the multipath signal in the CDMA receiver. In fact, what RAKE receivers do is to receive the signals of multipath signals through multiple correlation detectors and merge them together (Bottomley et al. 2000). A RAKE receiver is actually a classical diversity receiver designed for the CDMA system. The principle lies in that when the propagation delay exceeds a chip period, and multipath signals can actually be regarded as uncorrelated.

To guarantee that the spreading code generated in the receiver can synchronize with the spreading code in the received signal, time synchronization is needed. The correlator with delay-locked loop is a demodulation correlator with an early or later gate. The difference between early or later gate and demodulation correlator is +1/2 (or 1/4) chips, respectively. The subtraction of the correlation results of the early or later gate can be used to adjust the code phase. The performance of delay loop depends on the bandwidth of the loop.

For the RAKE receiver, each correlator can be regarded as a one finger with different delay. Figure 6.18 shows a RAKE combing model.

Since spread spectrum is more complicated in implementation than other diversity techniques, the spread spectrum with RAKE receiver is not used for diversity alone in practice. In systems like CDMA, spread spectrum signaling has already been used, and this kind of diversity can be applied naturally.

Since there are generally fast fading and noise in the channel, the phase of the received path changes greatly from the original transmitted signal, so the phase of the received channel should be rotated according to the result of channel estimation before combining. The channel estimation in the actual CDMA system is necessary. Pilot symbols carried in the transmitted signal are used to estimate the channel. According to whether there are continuous pilots in the transmitted signal, phase prediction methods based on continuous pilots and phase prediction methods based on decision feedback techniques can be used, respectively.

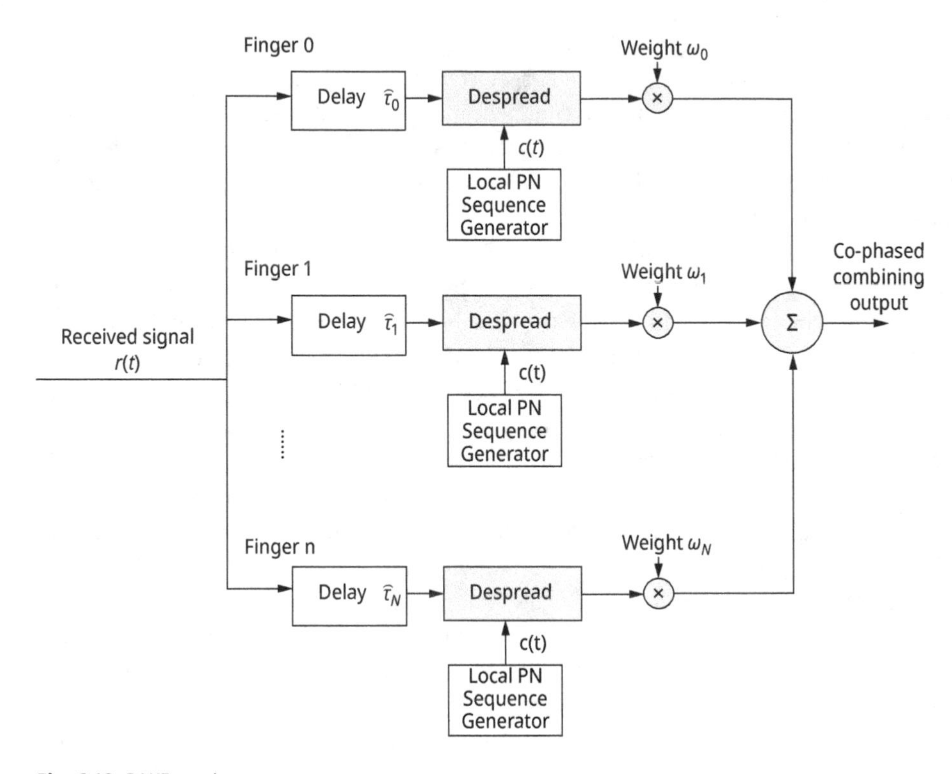

Fig. 6.18: RAKE receiver.

Problems

6.1 Given AWGN channel is used, assume the bandwidth of the transmitted signal is 8 kHz and the SNR is 3 dB, find the corresponding channel capacity. When the channel capacity is unchanged, assume the bandwidth is doubled and halved respectively, and calculate the signal power variation in these two cases.

6.2 A pseudo-random code rate is 5 Mc/s and the information rate is 16 kb/s. What are the radio frequency bandwidth and processing gain?

6.3 When the system is required to work in an environment where the interference signal is 250 times the desired signal, the output SNR is 10 dB and the internal loss of the system is 2 dB, what is the minimum processing gain of the system?

6.4 Given a 23-stage shift register, how long is the longest code sequence likely to be generated?

6.5 List the advantages and disadvantages of several pseudo-random codes discussed in this chapter, and figure out what kind of applications they can be applied to.

Chapter 7
Multicarrier modulation and multi-antenna systems

In Chapter 5, we discussed the main techniques to improve the link quality of wireless channels by compensating the fading channel impairments through independent paths, error detection coding, error correction coding and equalization. In Chapter 6, we discussed spread spectrum and RAKE receiver which improve performance of wireless communications without compensating the channel fading. In this chapter, we will discuss other two techniques which are very popular in modern wireless communications.

Multicarrier modulation is one technique to overcome ISI. In multicarrier systems, a high-speed data stream is divided into many low-speed substreams that are transmitted over different subchannels which can be less than the channel coherence bandwidth. The data rate of one such subchannel is much less than the total data rate so that the bandwidth of a subchannel is less than the coherent bandwidth of the channel. By this method, the signal in the each subchannel experiences flat fading instead of frequency-selective fading. If the same bit sequence over two or more subchannels is transmitted in multicarrier modulation, one kind of frequency diversity is implemented.

Communication systems with multiple antennas at transmitter and receiver are commonly called multiple input multiple output (MIMO) systems. The multiple antenna systems can dynamically increase the data rate or improve the performance through diversity.

In 5G mobile communication, massive MIMO has become the key technology because of the high data rate requirement. Its mechanism and key techniques of massive MIMO are different from the traditional MIMO due to the large number of antenna elements.

In this chapter, we will discuss multicarrier modulations and multiple antenna systems.

7.1 Multicarrier modulation

7.1.1 Overview of multicarrier modulations

Multicarrier modulation plays a very important role in modern wireless communications systems. However, its history dates back some 60 years (Weinstein 2009). In the late 1950s and early 1960s, it was first applied to military HF radio. In the 1950s, C. A. Doelz proposed a system called Kineplex. In this system, QPSK digital modulation

https://doi.org/10.1515/9783110751437-007

is adopted and 20 subcarriers are used. The spectrum of each subcarrier overlaps with each other and is orthogonal to each other, which greatly improves the spectrum utilization. The spacing of its adjacent subcarriers is approximately equal to the symbol rate.

In the 1960s, R.W. Hang et al. kept the subcarriers orthogonal by methods of filtering and limiting the bandwidth. In 1971, S. B. Einstein and P. M. Ebert used discrete Fourier transform (DFT) to carry out baseband modulation and demodulation of multiple carriers, which greatly reduced the system implementation complexity and made a great progress in the practical application of multicarrier modulation (Weinstein et al. 1971). This kind digital implementation of multicarrier modulation is called *orthogonal frequency division multiplexing* (OFDM). The idea of eliminating the interference between symbols by inserting guard intervals was proposed, but the interference would occur when the signal was no longer orthogonal among subcarriers after passing through a multipath channel (Weinstein et al. 1971). In 1980, A. Peled and A. Ruiz employed the cyclic prefix method to keep the orthogonal characteristics of the subcarriers after the signals pass through the multipath channel (Peled et al. 1980).

In 1985, multicarrier modulation was the first suggested to be used for wireless communications. From the 1990s, multicarrier modulation has found diverse applications that include digital audio and video broadcasting (DAB and DVB), digital subscriber lines (DSLs) using discrete multitone, the wireless local area networks (WLANs), fixed wireless broadband services, ultrawideband radios and the air interface in new-generation cellular systems.

Since OFDM is the most practical way to realize multicarrier modulation, we mainly discuss OFDM in this chapter.

7.1.2 Fundamentals of OFDM

As mentioned above, OFDM is an efficient digital implementation of multicarrier modulation (Cimini 1985), and it can also be regarded as a multiplexing technique. Multi-carrier transmission decomposes the data stream into several sub-bit streams, so that the bit rate of each sub-data stream is much lower. The low-rate multi-state symbols formed by such low-bit rate symbols are then used to modulate the corresponding sub-carriers, thus forming a parallel transmission system of multiple low-rate symbols. OFDM is an improvement on the traditional multicarrier modulation. In the traditional single-carrier frequency division multiplexing (FDM) system, the entire frequency band is divided into multiple non-overlapping sub-channels, and in order to avoid mutual interference, there are guard bands between each sub-channel. In this way, the FDM is achieved. The frequency spectrum of each subcarrier in OFDM system overlaps with each other, but remains orthogonal. The receiver can separate each subcarrier by coherent demodulation and eliminate the influence of intersymbol interference (ISI). As shown in Fig. 7.1, OFDM system can save more than half of the bandwidth compared with traditional single-carrier system.

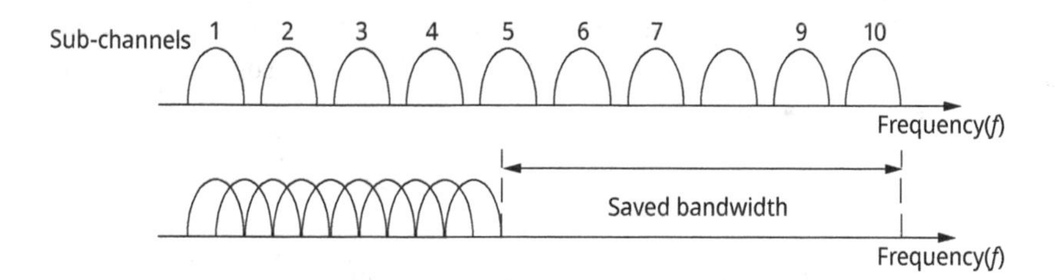

Fig. 7.1: Comparison of channel utilization between single carrier frequency division multiplexing system and OFDM system.

7.1.3 Principle of OFDM modulation and demodulation

OFDM signals can be expressed in the complex form as follows:

$$s(t) = \sum_{m=0}^{M-1} d_m(t) e^{j2\pi f_m t} \tag{7.1}$$

where $f_m = f_c + m\Delta f$ is the frequency of the mth subcarrier, f_0 is the 0th subcarrier frequency and $d_m(t)$ is a complex signal modulated on the mth subcarrier. $d_m(t)$ is a constant during a symbol time T_s, that is

$$d_m(t) = d_m \tag{7.2}$$

If the signal $s(t)$ is sampled at a sampling interval of T, then

$$s(kT) = \sum_{m=0}^{M-1} d_m e^{j2\pi f m kT} = \sum_{m=0}^{M-1} d_m e^{j(2\pi fc + m2\pi\Delta f)kT} \tag{7.3}$$

Assume that there are N sampling points during a symbol period T_s, that is:

$$T_s = NT \tag{7.4}$$

When OFDM is implemented in baseband, f_c can be set to be 0, eq. (7.3) can then be simplified as

$$s(kT) = \sum_{m=0}^{M-1} d_m e^{jm2\pi\Delta f kT} \tag{7.5}$$

Comparing the above equation with the form of discrete Fourier inverse transform (IDFT):

$$x(kT) = \sum_{m=0}^{M-1} X\left(\frac{m}{MT}\right) e^{j2\pi mk/M} \tag{7.6}$$

we can see that if $d_m(t)$ is regarded as a frequency-sampled signal, $s(kT)$ is the corresponding time domain signal. Comparing eqs. (7.6) and (7.5), we can conclude that the two equations are equal if we let

$$\Delta f = \frac{1}{NT} = \frac{1}{T_s} \tag{7.7}$$

Thus, if the carrier frequency interval $\Delta f = 1/T_s$ is chosen, the OFDM signal can not only keep the subcarriers orthogonal to each other but also be realized by DFT.

Since DFT is introduced to modulate the parallel data in OFDM system and $d_m(t)$ is constant during a symbol time Ts, its sub-band spectrum is $\sin(x)/x$ function. The spectrum structure of OFDM signal is shown in Fig. 7.2. The received OFDM signal can be demodulated by DFT at the receiver. Because the OFDM signal is realized through baseband processing, no oscillator group is needed, which greatly reduces the complexity of system implementation.

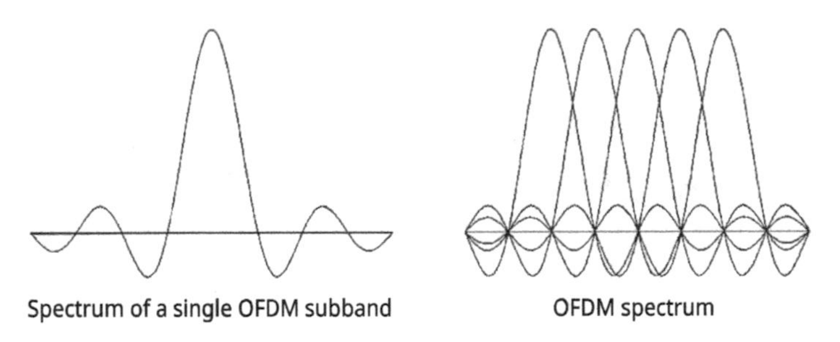

Fig. 7.2: Spectrum structure of OFDM signal.

7.1.4 Guard interval (GI) and cyclic prefix (CP)

In order to minimize ISI in OFDM systems, a guard interval is inserted between adjacent OFDM symbols. The duration of this guard interval T_g usually exceeds the maximum delay spread of the channel, so that the multipath component of the symbol will not interfere with the next symbol. The guard interval may contain no information; that is, the guard interval is an idle transmission period. However, in this case, due to the influence of multipath propagation, inter-carrier interference (ICI) occurs, that is to say, the orthogonality between subcarriers no longer hold, resulting in interference between different subcarriers. In order to eliminate ICI caused by multipath propagation, the OFDM symbol of length T can be periodically extended, and the guard interval can be filled with extended signals. The signal in the guard interval is called a cyclic prefix (CP) because this segment of signal is a copy of its corresponding OFDM signal in tail length of T_g. In the actual system, a CP needs to add to the OFDM signal before the OFDM signal is sent to the channel for transmission. At the receiver, the CP of each symbol should be removed before DFT demodulation. By adding CP to the symbol period of OFDM, the number of periods of the waveform contained in the delayed replica of the OFDM signal can be guaranteed to be an integer in the DFT period. In this way, the delay signal with delay less than the guard interval T_g will not generate ICI in the demodulation process. Figure 7.3 shows the time domain waveform diagram with CP added.

The effect of different length of protection interval on system performance is also different. Figure 7.4 shows the effect of different length of CP on system performance. In this simulation, a combination of 16QAM and OFDM is used. The maximum channel delay spread can be to 20 symbol time. It can be seen that the longer the CP is, the better the system performance is. The length of CP is larger than the maximum channel delay spread, and the CP length has no significant effect on the performance.

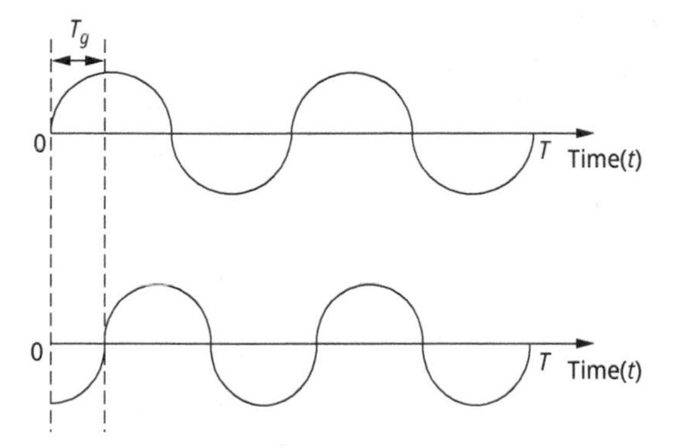

Fig. 7.3: Adding CP to an OFDM symbol.

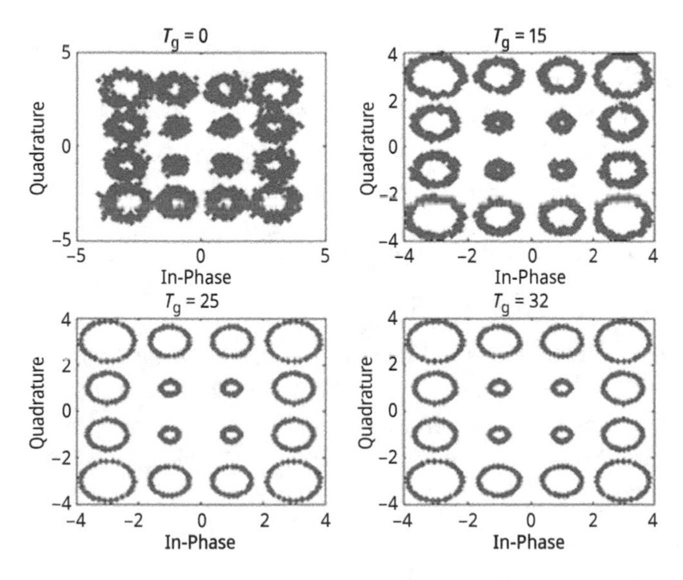

Fig. 7.4: Effect of different CP length on system performance.

7.1.5 Oversampling of IDFT

In practical system applications, N sampling points of an OFDM symbol or N output values from N-point IDFT often fail to truly reflect the changing characteristics of continuous OFDM symbols. Because if there is no oversampling, when these samples are sent to analog-to-digital converter (A/D), it is possible to generate pseudo-signal due to aliasing, which is not allowed in the system. The characteristics of this pseudo-signal depend on the sampling rate. When the sampling rate is less than twice the maximum frequency component of the signal, the signal will no longer contain the high-frequency component of the original signal when the sampling value is restored. Therefore, the signal shows a false low-frequency signal. To solve this problem, it is necessary to oversample OFDM symbols and add some sampling points between the original N sampling points to form pN sampling values. The implementation of oversampling is realized by subcarrier mapping. Before implementing IFFT operation, $(p-1)$ N zeros are added among the original input N values, and then pN-point IFFT is performed. Suppose a sequence of symbols in frequency domain $\{a_0, a_1, \ldots, a_{N-1}\}$ as the input N-point IFFT, now a more precise time-domain sampling points reflecting the continuous signal transformation can be obtained by four times oversampling. $3N$ zeros can be added in the middle to form $\{a_0, \ldots, a_{N/2-1}, 0, \ldots, 0, a_{N/2}, \ldots, a_{N-1}\}$, which is followed by $4N$-point IFFT. Thus, the oversampling of frequency domain signals can be realized, and the transformation of OFDM continuous symbols can be more accurately reflected. The Fourier transform of the time-domain sampled signal with T as sampling interval is composed of the repetition of the Fourier transform period of the continuous signal, whose repetition period is $1/T$. If the sampling interval is changed to T/P, the repetition period of the corresponding Fourier transform will be changed to P/T, and the spectrum width of the time-domain continuous signal needs to remain unchanged. Therefore, in frequency domain, it is equivalent to adding zero outside the bandwidth of the continuous signal, while in IFFT operation, it is equivalent to inserting zero between the frequency-domain data (sub-carriers).

7.1.6 Structure of OFDM modulation and demodulation

The generation of OFDM signal is based on fast Fourier transform (FFT), and its principle is shown in Fig. 7.5. The input binary bit sequence is first converted into parallel. According to the duration of the OFDM symbol T_s, $c_t = R_b{}^*T_s$ bits make up a group and the c_t bits are allocated to N sub-channels. Before the allocation, code mapping is needed to convert the bits into N complex symbol X_k, $k = 1, 2, \ldots, N$, where X_k denotes the symbol corresponding sub-channel k and represents b_k bits and

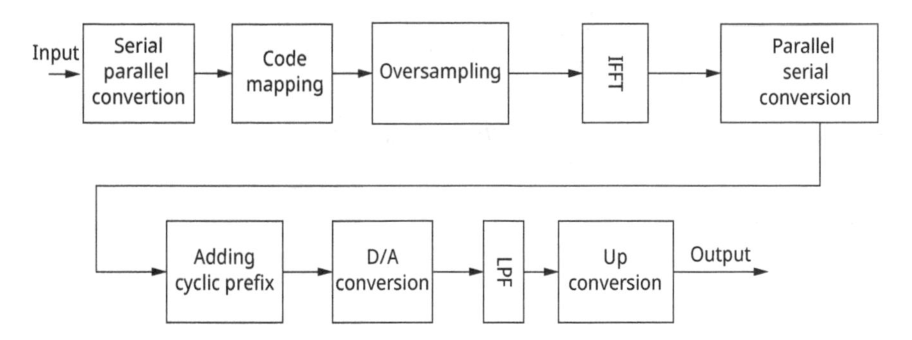

Fig. 7.5: The generation of OFDM signal.

$$c_t = \sum_{k=0}^{N-1} b_k \tag{7.8}$$

After zero-adding and oversampling, X_k is transformed into time-domain data by IFFT. At this point, parallel data needs to be converted into serial data, and then a cyclic prefix is inserted in front of each OFDM symbol. Finally, it is transmitted after digital/analog conversion, low-pass filtering and up-conversion.

The principle of the receiver of OFDM is shown in Fig. 7.6, and its processing is opposite to that of the transmitter.

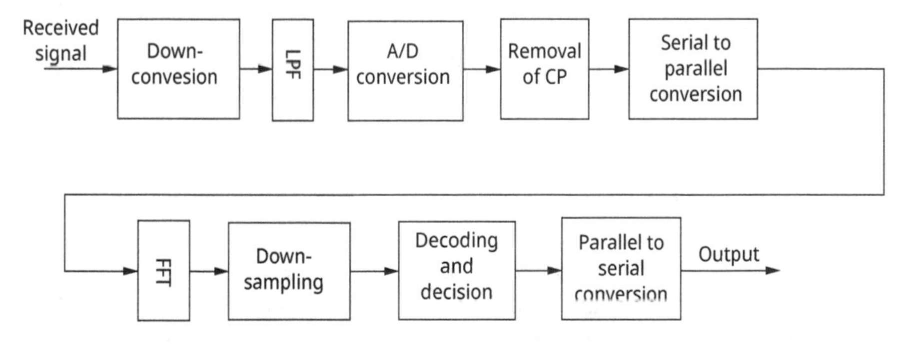

Fig. 7.6: Principle of OFDM signal receiver.

Because DFT is the baseband modulation in OFDM, it can be considered that the coding mapping of data is carried out in frequency domain and transmitted after being transformed to time domain signal by IFFT. In receiver, the frequency domain signal is recovered through FFT.

Different from the standard formulas of DFT and IDFT, DFT and IDFT in OFDM should satisfy the following relationship to keep the power unchanged before and after signal transformation:

$$X(k) = \frac{1}{\sqrt{N}} \sum_{n=0}^{N-1} x(n) \exp(-j2\pi nk/N) \tag{7.9}$$

$$x(n) = \frac{1}{\sqrt{N}} \sum_{k=0}^{N-1} X(k) \exp(j2\pi nk/N) \tag{7.10}$$

where $0 \leq k, n \leq N-1$.

7.1.7 Design of the parameters of OFDM

The first three parameters of an OFDM system are bandwidth, bit rate and CP. In general, CP is 2–4 times the $\delta_{Tm,}$ RMS value of channel delay spread. After estimating the maximum delay spread of the channel in which the system operates, the CP is determined. In practice, OFDM symbol period is generally chosen to about 5–6 times of CP. Equation (7.7) shows that the subcarrier spacing of the OFDM system is the reciprocal of the symbol period. Therefore, if the period of OFDM symbol is too long, the SNR loss caused by CP can be reduced, but this will make the subcarriers dense and increase the system complexity.

After determining the symbol period and CP, the number of bits that each OFDM symbol should transmit can be obtained from the required bit rate of the system and the duration of each OFDM symbol. Then the number of subcarriers is determined according to the coding scheme and bit rate. In this way, the number of subcarriers is not necessarily exactly integer power of 2, and the number of IFFT points is larger than the number of subcarriers and is integer power of 2 before IFFT.

7.1.8 Advantages and disadvantages of OFDM

OFDM technology has the following advantages:

1) By serial-to-parallel conversion of high-speed data streams, the length of symbols on each subcarrier is increased by N times, where N is the number of subcarriers, and thus the ISI caused by the multipath propagation of wireless channels would be effectively reduced. Apparently, this processing reduces the design complexity of the equalizer at the receiver. In some cases, even without the equalizer, the adverse effects of ISI can be eliminated by inserting the cyclic prefix.
2) In traditional FDM, the frequency band is divided into several non-overlapping sub-bands to transmit data stream at the same time, and sufficient protection bands between sub-channels must be preserved. Orthogonality also exists between subcarriers of OFDM system, which allows overlapping of subchannels. Therefore, compared with traditional FDM, OFDM can utilize the scarce spectrum resources more effectively.

3) The modulation and demodulation of OFDM can be implemented by FFT. The rapid development of large-scale integrated circuit and digital signal processor (DSP) makes this modulation and demodulation easy to implement.
4) Asymmetric high data rate transmission can be achieved by configuring different number of subcarriers in the uplink and downlink.
5) Each subchannel of OFDM system can use different coding methods. For subchannels with better channel conditions, some higher-order mapping methods, such as 64QAM, can be used. For deep fading subchannels, BPSK and other mapping methods may be used. This can make more effective use of the channel and improve the effectiveness of communication.

However, multiple orthogonal subcarriers exist in OFDM systems, and their output signals are sampled after the superposition of multiple subchannels. Therefore, compared with single carrier systems, the following disadvantages are given:
1) OFDM system is sensitive to frequency offset. Due to the overlapping spectrum of subcarriers in OFDM system, strict orthogonality between subcarriers is required. Because of the Doppler shift in wireless channel and the oscillator offset between transmitter and receiver, the orthogonality between subcarriers of OFDM system will no longer hold.
2) The peak-to-average power ratio of OFDM signal is larger. The output of multicarrier system is the superposition of multiple sub-channel signals. When multiple signals have identical phases, the superimposed signals will have much higher instantaneous power than the average power of the signals. This requires a high linearity of amplifier in transmitter, which may cause signal distortion and interference, deteriorate system performance and reduce the efficiency of low-frequency amplifier.

7.1.9 Applications of OFDM

As we have discussed, OFDM is a special multi-carrier digital modulation. The concept of OFDM originated in the mid-1950s. One of the main reasons for choosing OFDM is the capability of resisting frequency-selective fading and narrowband interference. In the 1960s, the concept of parallel data transmission and FDM was set up. The patent for OFDM was first published in January 1970.

As early as the 1960s, OFDM has been applied to a variety of high-frequency military systems, including KINEPLEX, ANDEFT and KNTHRYN. Taking KNTHRYN as an example, the variable rate data modem can use up to 34 parallel low-speed phase modulation subchannels with an interval of 82 Hz. Since then, Weinstein and Ebert have applied DFT to parallel transmission system to realize baseband modulation and demodulation, so that FDM can be realized through baseband processing in the process of FDM implementation, instead of using bandpass filter, subcarrier oscillator group and coherent demodulator.

Because there was no powerful integrated computing chip in the practical application at that time, the complexity of real-time Fourier transform, the stability of transmitter and receiver oscillator and the linear requirements of RF power amplifier all became the constraints of the implementation of OFDM. Until the mid-1980s, with the adoption of OFDM in digital audio broadcasting (DAB) in Europe (Sari et al. 1995), this method began to be concerned and widely used. In recent years, after the emergence of high-speed DSP chips, the superiority of OFDM has been recognized. With the combination of DSP and OFDM, Fourier transform/inverse transform, QAM, space-time block coding, lattice coding, soft decision and channel adaptation have gradually matured.

OFDM has found wide applications in DAB, DVB, IEEE802.11 standard-based WLAN and asymmetric high bit rate DSL based on existing copper twisted pairs in cable telephone network. Most of them make use of OFDM to effectively eliminate ISI caused by signal multipath propagation.

Another important application of OFDM is WLAN. In 1999, IEEE802.11a is adopted as a 5 GHz WLAN standard. The physical layer standard of this standard is OFDM. After that, 802.11 g operates at 2.4 GHz and 802.11 n operates at 2.4 GHz or 5 GHz adopted OFDM. The new high-efficiency wireless 802.11ax, also called Wi-Fi 6E, is a WLAN standard released in 2019, which also adopts OFDM. And of course, another key technique is MIMO which will be discussed later.

In the area of MAN, the Fixed Wireless Access Working Group of the OFDM Forum submitted a proposal to the MAN Committee of IEEE802.16.3 to adopt OFDM as the physical layer (PHY) standard of IEEE802.16.3 MAN in November 2000.

OFDM is also a key technique in 4G LTE and the 5G mobile communication system.

7.2 Multi-antenna systems

Multiple antennas at transmitter and receiver are capable of increasing the data rate through multiplexing or improving the system performance through diversity, and are commonly called MIMO systems (Paulraj et al. 2004). We have already covered diversity in Chapter 5. In MIMO systems, antennas in both transmitter and receiver sides can be used for diversity gain. In the 1980s and 1990s, the work of Winters (1987), Foschini (1996), Foschini et al. (1998) and Telatar (1995, 1999) predicated that the capacity of multiple antenna systems will be increased dramatically if channel side information (CSI) can be estimated accurately at receiver, and sometimes the CSI is also needed at transmitter. MIMO plays a key role in high-speed communications and finds wide applications in new-generation mobile communication (Tse et al. 2005). We will discuss the channel model, capacity and applications.

7.2.1 MIMO channel model

Assume that a MIMO system has N_T transmitting antennas and N_R receiving antennas. The channel model is shown in Fig. 7.7.

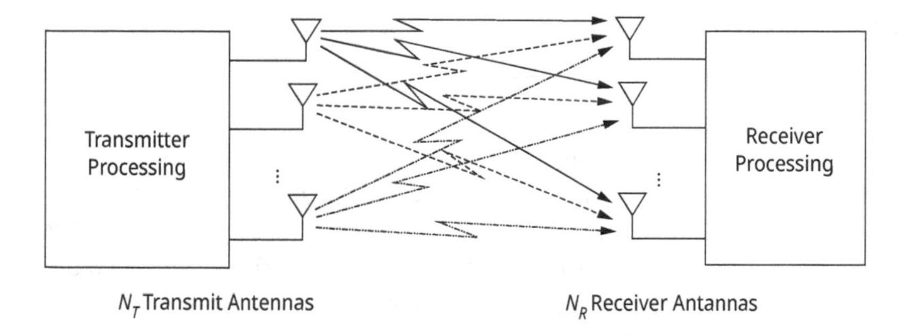

Fig. 7.7: MIMO channel model.

Generally, MIMO channel is modeled as Rayleigh fading, which is a typical scenario of NLOS in wireless communications. The signal path corresponding to each transmit antenna to receive antenna pair can be expressed as a complex gain factor. Let $h_{i,j}$ be the channel gain from transmitter antenna j ($j = 1, 2, \ldots, N_T$) to receive antenna i ($i = 1, 2, \ldots, N_R$), the MIMO channel model can be represented by a matrix $\boldsymbol{H}$:

$$\boldsymbol{H} = \begin{bmatrix} h_{1,1} & h_{1,2} & \cdots & h_{1,N_T} \\ h_{2,1} & h_{2,2} & \cdots & h_{2,N_T} \\ \vdots & \vdots & \ddots & \vdots \\ h_{N_R,1} & h_{N_R,2} & \cdots & h_{N_R,N_T} \end{bmatrix} \tag{7.11}$$

where each $h_{i,j}$ is identical and independent distributed complex Gaussian variable with zero mean unit variance; that is, the real part and the imaginary part are all Gaussian distributed with zero mean and 0.5 variance:

$$h_{i,j} = \mathcal{N}\left(0, \frac{1}{2}\right) + j\mathcal{N}\left(0, \frac{1}{2}\right) \tag{7.12}$$

The signal received by antenna i is expressed as

$$y_i = \sum_{j=1}^{N_T} h_{i,j} x_j + n_i \tag{7.13}$$

where x_j denotes the signal from transmit antenna j and n_i denotes the white Gaussian noise of zero mean. Let $\boldsymbol{x}$ and $\boldsymbol{y}$ denote the vector containing N_T transmit data and

N_R receive data, respectively. The above formula can be rewritten in the form of a matrix:

$$\boldsymbol{y} = \boldsymbol{H}\boldsymbol{x} + \boldsymbol{n} \tag{7.14}$$

According to singular value decomposition (SVD) theory, any matrix $\boldsymbol{H} \in \boldsymbol{C}_r^{N_R \times N_T} (r > 0)$ can be written as:

$$\boldsymbol{H} = \boldsymbol{U}\boldsymbol{D}\boldsymbol{V}^H \tag{7.15}$$

where $\boldsymbol{D}$ is the diagonal matrix of singular values of H with σ_i on the ith diagonal. $\boldsymbol{V}^H$ denotes the Hermitian of a matrix $\boldsymbol{V}$, and it is defined as its conjugate transpose of $\boldsymbol{V}$, i.e., $\boldsymbol{V}^H = (\boldsymbol{V}^*)^T$. The diagonal elements $\sigma_i (i = 1, 2, \ldots, r)$ are all nonzero singular values of matrix $\boldsymbol{H}$, and these elements are also the nonnegative square roots of eigenvalues of matrix $\boldsymbol{H}^H\boldsymbol{H}$. R_H of these singular values is nonzero, where R_H is the rank of the matrix $\boldsymbol{H}$. As we already know that the rank of a matrix should be less than or equal to the number of columns or rows, $R_H \leq \min(M_{NT}, M_{NR})$. $\boldsymbol{U}$ and $\boldsymbol{V}$ are unitary matrices of order N_R and N_T, respectively, so that $\boldsymbol{U}\boldsymbol{U}^H = \mathrm{I}_{\mathrm{N_R}}$ and $\boldsymbol{V}^H\boldsymbol{V} = \mathrm{I}_{\mathrm{N_T}}$. The row vector of $\boldsymbol{U}$ is the eigenvector of $\boldsymbol{H}\boldsymbol{H}^H$, and the column vector of $\boldsymbol{V}$ is the eigenvector of $\boldsymbol{H}\boldsymbol{H}^H$.

Substituting eqs. (7.15) into (7.14) yields

$$\boldsymbol{y} = \boldsymbol{U}\boldsymbol{D}\boldsymbol{V}^H\boldsymbol{x} + \boldsymbol{n} \tag{7.16}$$

Let $\tilde{\boldsymbol{y}} = \boldsymbol{U}^H\boldsymbol{y}$ which is called receiver shaping, $\tilde{\boldsymbol{x}} = \boldsymbol{V}^H\boldsymbol{x}$ which is called transmit precoding, and $\tilde{\boldsymbol{n}} = \boldsymbol{U}^H\boldsymbol{n}$, according to the property of unitary matrices $\boldsymbol{A}^H\boldsymbol{A} = \boldsymbol{I}$, we get

$$\begin{aligned}
\tilde{\boldsymbol{y}} &= \boldsymbol{U}^H(Hx + n) \\
&= \boldsymbol{U}^H(\boldsymbol{U}\boldsymbol{D}\boldsymbol{V}^H\boldsymbol{x} + \boldsymbol{n}) \\
&= \boldsymbol{U}^H(\boldsymbol{U}\boldsymbol{D}\boldsymbol{V}^H V\tilde{x} + \boldsymbol{n}) \\
&= \boldsymbol{U}^H\boldsymbol{U}\boldsymbol{D}\boldsymbol{V}^H V\tilde{x} + \boldsymbol{U}^H\boldsymbol{n} \\
&= \boldsymbol{D}\tilde{\boldsymbol{x}} + \tilde{\boldsymbol{n}}
\end{aligned} \tag{7.17}$$

Here the noise $\tilde{\boldsymbol{n}}$ has the same distribution as the noise $\boldsymbol{n}$, and $\|\tilde{\boldsymbol{x}}\|^2 = \|\boldsymbol{x}\|^2$, that is, the signal power remains unchanged. The channel is then equivalent to $r = \mathrm{rank}(\boldsymbol{H}) \leq \min(N_T, N_R)$ parallel subchannels:

$$\tilde{y}_i = \sigma_i\tilde{x}_i + \tilde{n}_i, \; i = 1, 2, \ldots, r \tag{7.18}$$

The singular value of channel matrix $\sigma_i (i = 1, 2, \ldots, r)$ can be regarded as the channel gain of the subchannels, and they are also nonnegative square roots of eigenvalues of matrix $\boldsymbol{H}^H\boldsymbol{H}$, $\lambda_1 \geq \lambda_2 \geq \cdots \geq \lambda_r > \lambda_{r+1} = \cdots = \lambda_{N_T} = 0$. By SVD, MIMO channel becomes equivalent parallel subchannels, which can be shown in Fig. 7.8.

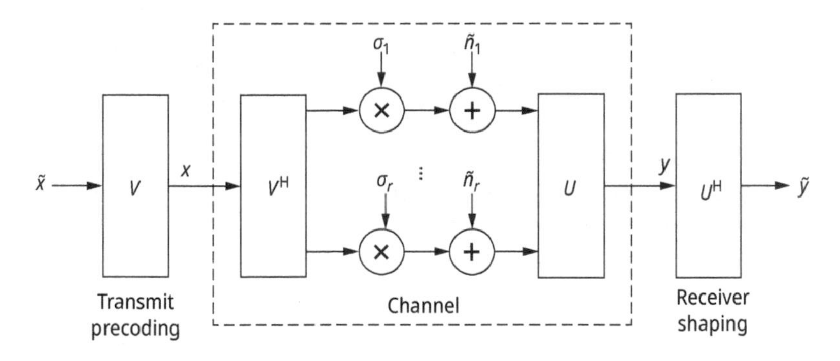

Fig. 7.8: Equivalent parallel subchannels of MIMO.

It can be seen that MIMO provides space resources for wireless communication systems in addition to time domain and frequency domain resources, which provides greater flexibility for system design.

7.2.2 MIMO channel capacity

As discussed above, by SVD, a MIMO channel can be equivalent to r parallel subchannels. According to the channel capacity of parallel Gaussian channel, MIMO channel capacity can be easily obtained. Let's assume $\boldsymbol{x} \sim \mathcal{CN}(0, \boldsymbol{Q})$, $\boldsymbol{n} \sim \mathcal{CN}(0, \sigma_n^2 \boldsymbol{I})$ and the total transmit power constraint is P_T, that is $\text{tr}(\boldsymbol{Q}) \leq P_T$. For a given channel $\boldsymbol{H}$, the capacity is

$$C = \sum_i \log_2 \left(1 + \frac{P_i \lambda_i}{\sigma_n^2}\right) \tag{7.19}$$

where $\sum_i P_i \leq P_T$.

7.2.2.1 CSI known at receiver

When only the receiver knows the channel side information (CSI) and the transmitter does not, the maximum channel capacity is obtained by averaging the power allocation on the transmitting antenna, i.e.,

$$P_i = \frac{P_T}{N_T} \tag{7.20}$$

7.2.2.2 CSI known at receiver and transmitter

When the transmitter knows the CSI, the maximum channel capacity and the corresponding power allocation policy can be obtained by the water-filling algorithm:

$$P_i = \begin{cases} \mu - \frac{\sigma_n^2}{\lambda_i} & \text{when } \mu \geq \frac{\sigma_n^2}{\lambda_i} \\ 0 & \text{otherwise} \end{cases} \tag{7.21}$$

where λ_i is the nonnegative square root of the eigenvalue of matrix $\boldsymbol{H}^H\boldsymbol{H}$, and μ is the cutoff value calculated from the water-filling algorithm.

The principle water-filling algorithm is shown in Fig. 7.9.

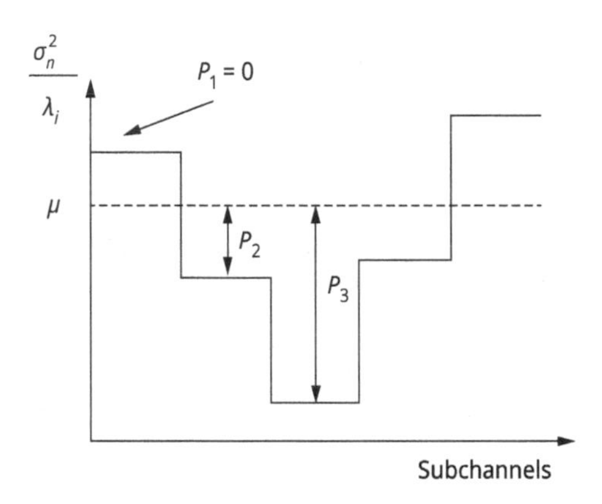

Fig. 7.9: Water-filling power allocation policy of MIMO parallel subchannels.

7.2.3 Applications of MIMO

MIMO is an important breakthrough in modern communications. On one hand, it can greatly improve the system capacity and spectrum utilization, and not at the cost of time and frequency resources; on the other hand, it transforms the multipath propagation that is unfavorable to wireless transmission into a favorable factor, and effectively improves the performance of communication systems at different metrics by using multipath effectively. The two main applications of MIMO system are spatial multiplexing and spatial diversity.

7.2.3.1 Spatial multiplexing

Spatial multiplexing refers to transmitting different data streams on different antennas, making full use of spatial characteristics and aiming at improving the data rate in the channel under a certain bit error rate. A typical example is the BLAST (Bell Laboratories Layered Space Time) proposed by Bell Laboratory. In the BLAST system, the signals that need to be transmitted are firstly converted into multiple parallel signals. The signal is transmitted by different antennas at the same time and the same frequency band, and the receivers distinguish their data streams by using the difference of spatial parallel subchannels. According to the transmission order of encoded data stream, BLAST can be divided into three types: horizontal layered space-time code (H-BLAST) (Li et al. 2005), vertical layered space-time code (V-BLAST) (Wolniansky et al. 1998) and diagonal layered space-time code (D-BLAST) (Prasad et al. 2002), where the details will not be given here.

7.2.3.2 Spatial diversity

Spatial diversity is to increase the reliability of communication by using spatial redundancy to transmit data containing the same information on a parallel channel in space. The space-time trellis code (STTC) proposed by Tarokh of AT&T Research Institute can get both the diversity gain and coding gain simultaneously, but its decoding complexity increases exponentially with the number of antennas and data rates. The other kind of space-time block code (STBC), which uses orthogonal coding, can be decoded just by simple linear processing, and the complexity is lower. Cadence's Alamouti first proposed a space-time block code for two transmitting antennas, which was later extended to orthogonal space-time block codes for more than two antennas. We will show two-antenna Alamouti scheme as an example.

Let's first make an assumption that the channel gain is constant over two symbol periods. Two different symbols s_1 and s_2 are sent simultaneously over the first symbol period each with energy $E_s/2$ from antennas 1 and 2, respectively. And then two other symbols are sent from antenna 1 and antenna 2 over the second symbol time, which are $-s_2{}^*$ and $s_1{}^*$ each with a symbol energy of $E_s/2$.

Let $h_i = r_i e^{j\theta_i}$, $i = 1, 2$ denote the complex channel gains between the ith transmit antenna and the receive antenna. In the first symbol time, the received symbol is given

$$y_1 = h_1 s_1 + h_2 s_2 + n_1 \tag{7.22}$$

where n_1 is the additive white Gaussian noise (AWGN) received at the first symbol time.

Similarly, the received symbol in the second symbol period is

$$y_2 = -h_1 s_2^* + h_2 s_1^* + n_2 \tag{7.23}$$

where n_2 is the corresponding AWGN.

Suppose that the AWGN is zero mean and the power is N. We form the sequentially received symbols y_1 and y_2 into a vector $\mathbf{y} = [y_1\, y_2]^{\mathrm{T}}$, then

$$\mathbf{y} = \begin{bmatrix} h_1 & h_2 \\ h_2^* & -h_1^* \end{bmatrix} \begin{bmatrix} s_1 \\ s_2 \end{bmatrix} + \begin{bmatrix} n_1 \\ n_2^* \end{bmatrix} = H_A s + n \tag{7.24}$$

where

$$\mathbf{s} = \begin{bmatrix} s_1 \\ s_2 \end{bmatrix} \tag{7.25}$$

$$\boldsymbol{n} = \begin{bmatrix} n_1 \\ n_2^* \end{bmatrix} \tag{7.26}$$

and

$$\boldsymbol{H}_A = \begin{bmatrix} h_1 & h_2 \\ h_2^* & -h_1^* \end{bmatrix} \tag{7.27}$$

Let $\boldsymbol{A}^H$ denote the Hermitian of a matrix $\boldsymbol{A}$, and it is defined as its conjugate transpose of $\boldsymbol{A}$

$$\boldsymbol{A}^H = (\boldsymbol{A}^*)^T \tag{7.28}$$

From the structure of $\boldsymbol{H}_A$, we know that

$$\boldsymbol{H}_A^H \boldsymbol{H}_A = (|h_1{}^2| + |h_2{}^2|)\boldsymbol{I}_2 \tag{7.29}$$

is diagonal. By left multiplying $\boldsymbol{y}$ by $\boldsymbol{H}_A^H$, we get a new vector z as follows:

$$\boldsymbol{z} = \boldsymbol{H}_A^H \boldsymbol{y} \tag{7.30}$$

Substituting eqs. (7.29) and (7.24) into (7.30), we get:

$$z = [z_1 \quad z_2]^T = (|h_1^2| + |h_2^2|)\boldsymbol{I}_2 \boldsymbol{s} + \tilde{\boldsymbol{n}} \tag{7.31}$$

where $\tilde{\boldsymbol{n}} = \boldsymbol{H}_A^H n$ is a zero mean complex Gaussian noise vector.

Thus, the two symbol transmissions are effectively decoupled into two components, and each corresponds to one of the transmitted symbols:

$$z_i = (|h_1^2| + |h_2^2|)s_i + \tilde{n}_i, \; i = 1, 2 \tag{7.32}$$

This equation shows that the Alamouti scheme gets a diversity gain of 2, which is the full diversity gain with two antennas, even though the transmitter does not know CSI.

The received SNR for z_i is

$$\gamma_i = \frac{(|h_1^2| + |h_2^2|)}{2N_0} E_s \tag{7.33}$$

The SNR is the half of the sum of the SNRs on each branch because the transmitted signal energy on each symbol time is half of the total symbol energy. Equation (7.33) shows that the array of Alamouti scheme is only 1; it denotes that if there is no fading, the reserved SNR in this scenario is not increased. The generalized cases with more than two antennas can be found in Paulraj et al. (2003).

7.2.4 Massive MIMO

As stated above, massive MIMO is a key technology in 5G mobile communications because of its supporting for high data rate transmissions. The mechanism and key techniques of massive MIMO are different from that of the traditional MIMO. We will discuss the fundamental of massive MIMO in this subsection.

7.2.4.1 Concept of massive MIMO

Massive MIMO is a physical layer technology for wireless access which is based on MIMO. It uses large antenna arrays and to simultaneously serve many autonomous terminals by using the principle of space division multiplexing and beam forming. Large antenna arrays can support dozens of independent spatial data streams by increasing the number of antennas on the basis of existing multi-antennas, which will multiply the spectrum efficiency of multi-user systems and play an important role in supporting the capacity and rate requirements of 5G systems. Besides the high spectral efficiency, massive MIMO also provide high energy efficiency, by virtue of the array gain, which allows low radiated power.

Different from the traditional MIMO technology, when the number of antenna elements tends to be very large (infinite), the channels of massive MIMO tend to be orthogonal. The performances of the system are only related to large scale, but not to small scale. Pilot design is always used in traditional MIMO; however, in massive MIMO, the pilot design of hundreds of antennas in base stations consumes a lot of time-frequency resources, and the pilot-based channel estimation is not desirable. TDD can use the reciprocity of the channel to estimate the channel, and it does not need pilots to estimate the channel.

7.2.4.2 Characteristics of massive MIMO in 5G communication networks

Massive MIMO system has many new features compared with the traditional MIMO system because the number of antennas installed in the base station increases by tens or hundreds of times:

(1) The randomness of channel is decreased and the certainty is increased. In traditional MIMO systems, because of the small number of antennas, each channel formed by transmitter and receiver has certain independence and individual characteristics. Furthermore, the continuity between different channels is poor. However, the number of antennas in Massive MIMO system is fairly large, and the previous independent channels become a channel matrix, so that the individual characteristics form a certain certainty in the matrix. Meanwhile, the matrix can be used to decompose the whole operation in some way, so as to reduce the difficulty of operation. To improve the accuracy, the aperture of the antenna matrix can be increased.

(2) The anti-interference ability is enhanced. Since there are a large number of antennas in the base station layout, thermal noise, interval interference and other interference conditions will cause little interference to the communication system. At the same time, the hundreds of antennas can further enhance the useful power of the signal. The coordination dependence of different base stations is reduced, so the computational complexity is greatly reduced.

7.2.4.3 Technical challenges of massive MIMO

(1) Resource scheduling:
Massive MIMO is used for communication transmission, so it is necessary to pair the base station antennas to form multiple transmission channels. It involves the problem of antenna selection and users grouping. Due to the simultaneous presence of multiple users in the communication system, the effective allocation of spectrum resources becomes a key problem.

(2) The accuracy of channel information and the stability of the algorithm:
The economic benefits brought by massive MIMO depend largely on the accuracy of channel information estimation and the time delay spread of the channel, which requires effective improvement of information accuracy and reduction of feedback overhead in practical research.

Because of the complex application environment, the limited speed of information feedback and many unavoidable signal interference factors, the algorithm is generally unstable, which seriously affects the advantages of massive MIMO technology.

(3) Pilot contamination. Pilot contamination is mainly due to the use of the same pilot sequence in near cells.
This reuse of pilot is the cause of contamination. Of course, pilot contamination is not a new problem of massive MIMO, but it is undeniable that this problem became more serious in massive MIMO. In the aspect of how to deal with pilot contamination, researches have given some results, and the method of shifting pilot sequence can be used. The principle of the method is that the pilots in different positions in the frame are used in adjacent cells. In this way, although it seems that the same pilot sequence is used, it is actually different, so that even in near cells, multiple users can employ the pilot simultaneously, there will be no pilot contamination problem.

Problems

7.1 Describe the main difference OFDM and FDM.
7.2 What are the main advantages of OFDM? What are the main drawbacks of OFDM?

7.3 Show that the subchannels are still orthogonal when we use CP that is greater than the delay spread. How about if we use an all-zero sequence to substitute the CP?
7.4 Explain the principle why the capacity of a MIMO system can be increased.
7.5 Why the spatial diversity of multiple antennas at transmitter has a diversity gain but does not have an array gain?
7.6 Figure out the main difference between the traditional MIMO and massive MIMO.

Chapter 8
Duplexing, multiple access and cellular design fundamentals

After the discussion on the wireless channel, the performance of digital modulations over wireless channels and techniques to improve the performance of a wireless link, we now discuss some related techniques in wireless cellular design. The first technique in air interface is the duplexing, which is used to separate the uplink (mobile station (MS) to base station (BS)) and downlink (BS to MS) transmission. The second technique is the multiple access technique, which is used to separate different users. We will also discuss design fundamentals in the cellular system.

8.1 Duplexing

In a cellular system, duplexing is the key technique in air interface to guarantee the bidirectional transmission between downlink and uplink.

Basically, there are two fundamental duplexing techniques adopted in cellular networks: time domain duplexing (TDD) and frequency domain duplexing (FDD).

8.1.1 FDD

In FDD, the uplink and downlink data are transmitted in different frequency bands, which are shown in Fig. 8.1.

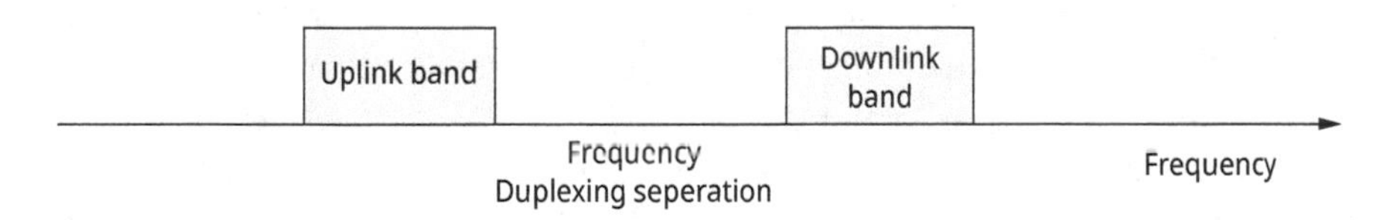

Fig. 8.1: Diagram of FDD.

FDD is very popular in cellular systems. For example, the most successful mobile communication system worldwide – the Global System for Mobile (GSM) communications – adopts FDD. There are three versions of GSM: the original version, GSM900, frequency band from 890 to 915 MHz is used for the uplink (connection from the MS to the BS) and frequency band from 935 to 960 MHz is used for the downlink. Although the frequency separation of uplink and downlink is 20 MHz, the frequency spacing between the uplink and downlink for any given connection is 45 MHz.

https://doi.org/10.1515/9783110751437-008

For the second version GSM–GSM1800, the uplink band is 1710–1785 MHz, and the downlink band is 1805–1880 MHz.

In North America 1900 MHz version, frequency band 1850–1910 MHz is used for the uplink and the band 1930–1990 MHz is used for the downlink.

The advantage of FDD is that it is sufficient for relatively cheap duplex filters to achieve very good separation between the uplink and downlink.

8.1.2 TDD

In TDD, uplink data and downlink data are sent at different time slots, which are shown in Fig. 8.2.

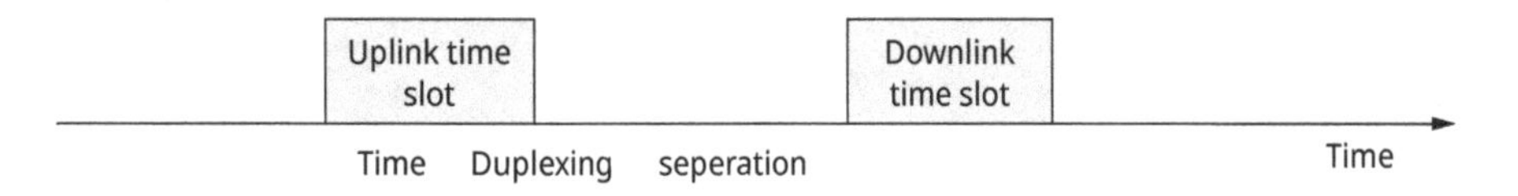

Fig. 8.2: Diagram of TDD.

TDD can be used in conjunction with time division multiple access (TDMA), code division multiple access (CDMA) or packet radio (PR) networks. If TDD is used in conjunction with TDMA, the time slot number is generally $2N$, where each of the N users is assigned two timeslots, one for the uplink and one for the downlink.

One advantage of TDD over FDD is that we can flexibly allocate the number of uplink and downlink slots, which is very useful for unbalanced applications. For example, in Internet applications, downlink data rates are often greater than uplink data rates.

Another advantage of TDD is that the same frequency band is used in the uplink and downlink channels and the duplexing time is much smaller than the coherence time. We can take it for granted that the uplink and downlink channels are reciprocal so that the received downlink channel side information (CSI) can be used as the CSI of the uplink channel, thus eliminating the complexity of channel feedback. This principle can also be applied to downlink. In the TDD mode, the implementation of adaptive modulation is simplified, and various smart antenna systems are easy to implement.

One problem of TDD is that its efficiency is limited by the time delay or cell coverage. Suppose that an MS at a cell boundary where the distance to BS is d and the corresponding runtime is d/c, where c is the velocity of light, transmits a sequence of data which is length of a time slot T to BS, the data will arrive at BS $T + d/c$. Only after the data is received, BS starts to transmit a time slot data of length T. This data will start reaching MS at time $T + 2d/c$ and this time slot transmission finishes at $2T + 2d/c$. In other words, if MS starts its transmission at time 0, it starts to get echo at $T + 2d/c$, and the MS has experienced a dead time of $2d/c$. If the cell size is large, the efficiency loss would be considerable.

8.1.3 Combination of FDD and TDD

In the case of semi-duplex systems, a combination of FDD and TDD may be adopted, where the uplink and downlink are distinguished by both time and frequency. When FDD is used for simultaneous transmission, it generally requires high-quality duplex filters. But when uplink and downlink of FDD occurs at a different time slot, the duplex filters do not have to be good enough. For example, in GSM, the uplink and downlink transmission is at different frequency bands and the time is also different by two timeslots.

8.1.4 Full duplex (FD)

FDD and TDD can offer transmission and reception for MS; however, they are basically half duplex because they use frequency division or time division to avoid self-interference.

Full duplex (FD) systems are capable to transmit and receive on the same spectrum simultaneously with efficient self-interference (SI) cancellation techniques (Zheng et al. 2013). As a result, FD can achieve almost double spectrum efficiency with comparison to FDD or TDD.

FD is a hot topic in modern wireless communications, which may be used in relay systems with channel estimation and multiple input multiple output (MIMO) technologies. The details will not be treated here; interested readers may refer to Liu et al. (2015), Goyal et al. (2015) and Sabharwal et al. (2014).

8.2 Multiple access

8.2.1 Introduction to multiple access

Multiple access techniques solve the problem on how to communicate with many users simultaneously. In this chapter, we assume that one BS communicates with multiple MSs.

There are three major multiple access techniques adopted in wireless network systems. By frequency division multiple access (FDMA), different users are assigned different frequency bands to transmit information. By TDMA, different users are assigned different timeslots to transmit information. By CDMA, different users are assigned different codes to transmit information. Besides the three main methods, there also other methods, such as space division multiple access, PR and the combination of two multiple access methods.

8.2.2 Frequency division multiple access (FDMA)

Conceptually, FDMA is quite simple. It assigns an individual frequency subband or channel to one user, which is shown in Fig. 8.3. The channels or subbands are the recourse and a specific channel is generally allocated to connection during the connection setup period, so the connection will keep this resource until the end of it.

In FDMA systems, like the first-generation cellar system, it is always combined with FDD, two channels are assigned to one user, one for downlink and one for uplink, and these two channels have a fixed duplex distance.

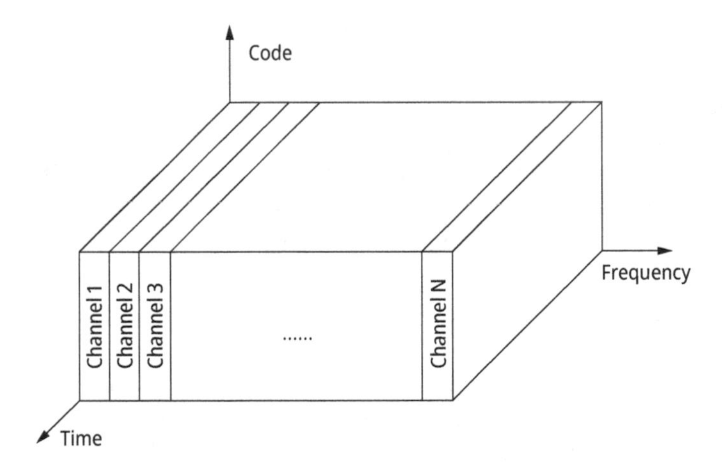

Fig. 8.3: Diagram of FDMA.

FDMA is easy to implement since it is done in analogy mode; its synchronization can be easily established during the connection setup period and then tracked during the connection by simple algorithm. However, when FDMA is used in speech communications, its spectral efficiency is limited. Even when an FDMA channel is not in use, it cannot be used by other users.

The bandwidth of each FDMA subchannel is relatively narrow (≤30 kHz), which is less than the channel coherence bandwidth; thus, the channel is a flat fading channel and no serious inter-symbol interference occurs. That's why equalization is not required. However, the narrow bandwidth also has drawbacks, and the carrier frequency synchronization will be sensitive to the frequency offset, which requires that the local oscillators must be very accurate and stable. The frequency jitters can also cause inter-channel interference (ICI). To overcome this problem, either accurate oscillators and steep filters or guard bands are required. They may cause high cost or low spectral efficiency. The narrow bandwidth also makes FDMA sensitive to frequency modulation. Since each subband is narrow, the BS generally uses a large number of subchannels. Intermodulation will occur when these signals are amplified by the same power amplifier, and undesirable third-order modulation products will be created. If separate amplifiers are designed for each of the channels or highly linear amplifiers for the combined total signal, the resulting BS will be costly.

Due to the above properties, the applications of FDMA are limited. Early FDMA was only used in analog systems, such as the first-generation cellular system. Nowadays, FDMA is generally combined with other multiple access methods, such as MF-TDMA or MF-CDMA. In these methods, the frequency is firstly divided into subbands, and in each subband, multiple time slots or codes are assigned to different users. FDMA can also be used in high-speed applications so that each subband is wide enough and only a few number of subbands in total; thus, the drawbacks can be avoided.

8.2.3 Time division multiple access (TDMA)

In TDMA systems, the spectrum is divided into time slots, and different users are assigned different ones. Generally, a user is assigned a pair of time slots, one for uplink and one for downlink. The principle of TDMA is shown in Fig. 8.4.

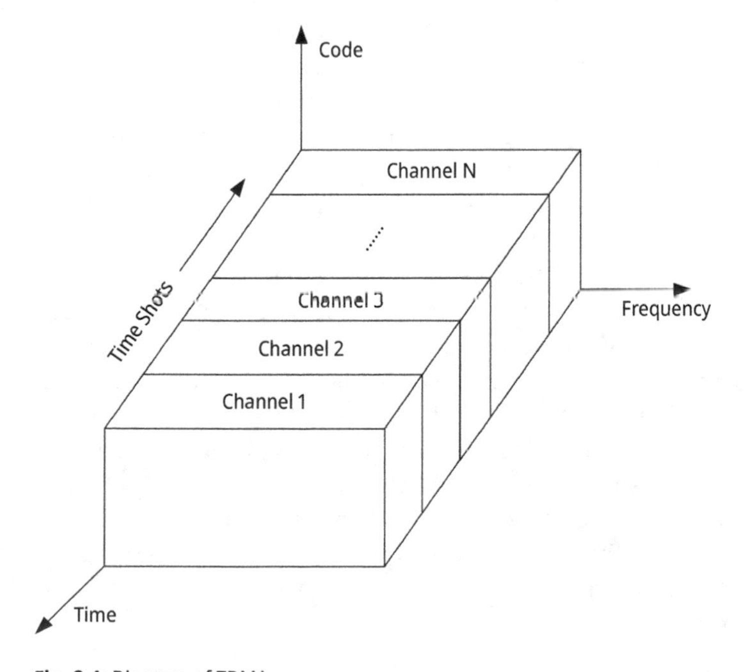

Fig. 8.4: Diagram of TDMA.

The time axis is divided into cyclic time unit with fixed duration. Suppose there are N users, each time unit is then subdivided into N timeslots of fixed duration, and each user occupies a cyclically repeating time slot. We can define the time unit of N time slots in a frame, and each user occupies the same time slot in the periodically occurring frames. This means that only $1/N$ of the total time can be used for a user. TDMA transmits data in a buffer-and-burst mode. During the assigned time slot, the user has to transmit buffered data of N time slots within one time slot with a high data rate by using the whole system bandwidth. It will keep silence in the other N–1 time slots, which are occupied by the other N–1 users. At the receiver, the high-speed data is then converted into low-speed data. FDD and TDD can be combined with TDMA. In TDMA/TDD, half of time slots are used for uplink and half of the slots are used for downlink. In TDMA/FDD, all the slots will be used either for uplink or downlink, but the carrier frequency is different. In practice, the time slots between uplink and downlink generally have several time slots with time difference so that duplexers are not needed in MS. A diagram of frame structure is shown in Fig. 8.5. The preamble sequence is the control information including address and synchronization sequence to let the BS and MS to identify each other. Trail bits and guard bits are used for synchronization between frames and slots, and the domain names and lengths are different in different wireless standards.

As stated above, TDMA transmits information in the burst mode, so that it occupies a higher bandwidth but only works at $1/N$ of the total time, where N is the number of users. In this way, the transmitter of the user can be turned off when not in use. Since the transmission does not work in most of the time, the *handoff* would be much simpler than that of FDMA system since the MS has enough time to listen to other BSs in its idle time slots. However, the high bandwidth may be greater than the channel coherence bandwidth, so frequency-selective fading may occur, and techniques like equalizers are required to overcome this problem. Frequency diversity may be exploited to improve the performance.

Guard bits are necessary for TDMA slots. In Section 8.1, we have analyzed the length of guard time in TDD. In TDMA/FDD mode, the guard length between two time slots need to consider the time difference between the largest runtime of MS at the cell boundary and the smallest runtime of MS near the BS. Since TDMA works at the burst mode, the synchronization is needed for each of the time slot. If there is time difference between the transmission of current burst and the previous burst of the user, the channel estimation has to be done for each of the burst. Of course, in the TDMA/TDD mode, reciprocity can be used to improve the efficiency of channel estimation.

One advantage of TDMA is that different users can be allocated to different number of slots according to their service types and quality of service requirements.

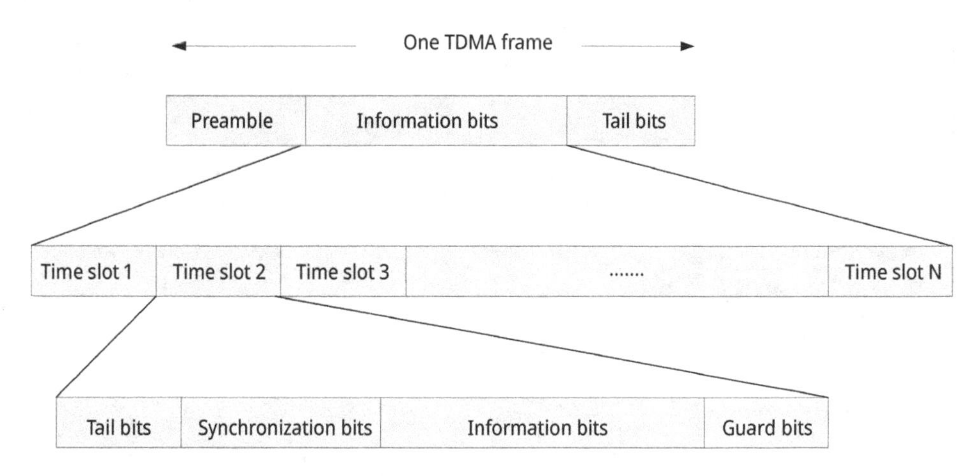

Fig. 8.5: Diagram of TDMA frame and time slot.

8.2.4 Code division multiple access (CDMA)

The CDMA principle is based on the direct spectrum spread system we have discussed in Chapter 6. By multiplying a very large bandwidth pseudo-noise (PN) code signal which is called a spreading signal, the narrowband user information is expanded into a large bandwidth of the CDMA system The PN coder has a chip rate that is an integral multiple of the information rate. Each user has its own PN code and this specific PN code of each user is approximately orthogonal to each other, so the information can be transmitted in the same carrier frequency simultaneously. The diagram of CDMA is shown in Fig. 8.6. At the receiver, correlation with local PN codes is performed to pick the desired codeword. Because of orthogonality, the correlation output with other undesired codewords can be regarded as noise.

In CDMA system, the signal of all users is overlapped in time and frequency domain, and the power of other users at the receiver determines the noise of the desired one. The power of each user must be controlled so that the power received at the BS at approximately the same level or the *near-far effect* will occur.

In CDMA, if all MSs send signal with approximately the same power, at the receiver, the signal from those at the boundary of the cell which is far from BS would be much lower than the signal from those at the center of the cell which is near to the BS. The stronger received signal will increase the noise level of the weaker received signal and make the demodulation of weaker received signal more difficult. This effect is *the near-far* problem. Channel inversion is a typical method to solve this problem, which uses the power control to let the signal reach the BS with an approximately same power. Power control is implemented according to the radio signal strength indicator (RSSI) level of each of the MS measured at BS and sent to the power control command in the downlink control channel.

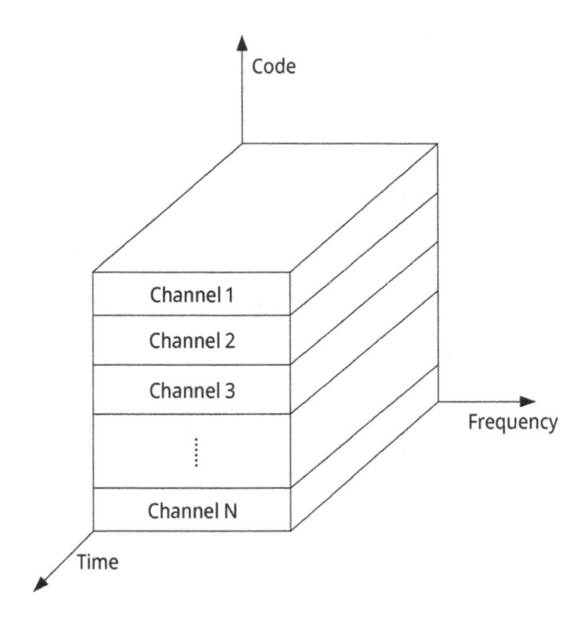

Fig. 8.6: Diagram of CDMA.

A unique property of CDMA is that it has soft capacity limit so that we can increase the number of users when necessary if enough number of orthogonal PN codes exist. Of course, the larger number of users, the higher the noise floor of the desired signal in demodulation will be. As a result, the performance will be improved with the number of users decreased, and the performance will be degraded with the number of users increased; thus, there is no hard limit on the number of users. In order to improve the performance of CDMA, multi-user detection (MUD) algorithm can be applied. Especially when the spreading sequences of different users are not exactly orthogonal, the signal of other users makes non-zero contributions to the desired user in the decision of the receiver, which is called self-jamming. However, MUD algorithm generally has a high complexity and its application is not very popular. If the CDMA users are synchronized, the complexity will be much lower. That is why in MUD algorithm, TD_SCDMA is used.

In CDMA system, the chip rate is much higher than the information data rate, and the signal bandwidth is also generally greater than the channel coherence bandwidth. The inherent frequency diversity mitigates the effect of multi-path effect, and the spectrum spread principle of CDMA has the capacity of narrowband interference suppression. As stated in Chapter 7, a RAKE receiver can be designed to improve the reception by collecting different copies of signal with different time delays.

In CDMA system, co-channel cells are used, and this mechanism allows soft handoff by using macroscopic spatial diversity. The mobile switch center (MSC) monitors one particular user simultaneously from two or more BSs and may pick the best one of all the signals at any time without switching.

8.2.5 Multiple access for packet radio (PR)

Different from the access methods for cellular phone network, PR does not allocate a dedicated channel for each user. In PR, the data are broken down into packets, and each of the packets is transmitted over the channel independently without coordination. The data traffic from each user is in the burst mode, and the simultaneously transmitted packets may collide. The BS receiver is responsible for detecting the collision, and an ACK (acknowledgment) or NACK (negative acknowledgment) is broadcast by the BS to inform the desired user (or all users) whether the previous transmission is successful or not. If the transmitter did not get an ACK or NACK in a predetermined time, it also assumes the transmission as a failure and will transmit the packet again. The second difference of the multiple access of PR is that each packet may be routed to its destination via independent relay stations, and each wireless device may act as a relay node for information originating from another wireless node.

The major multiple access methods for PR are ALOHA, carrier-sense multiple access (CSMA) and packet reservation (polling). Let us discuss the fundamentals of them.

8.2.5.1 ALOHA

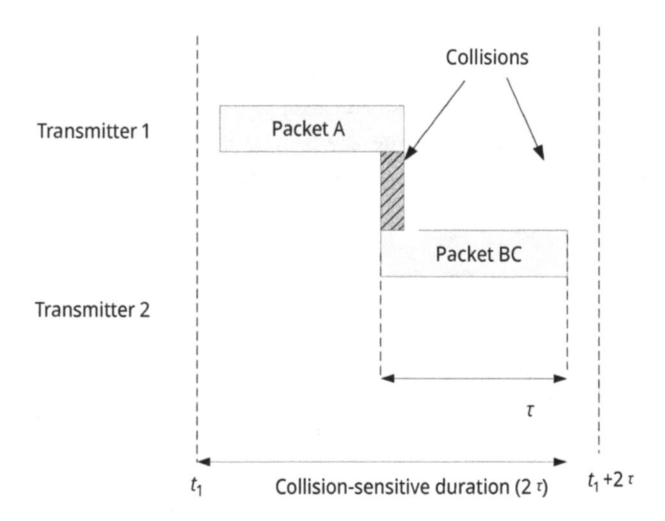

Fig. 8.7: Occurrence of collisions in ALOHA system.

The ALOHA protocol was proposed and implemented in first packet-based wireless radio system, which was developed at the University of Hawaii in 1971. This wireless computer network provided wireless connections of computers at seven campuses spread out over four islands through a central computer on Oahu. The multiple access method is called ALOHA, and the underlying principles are still used today in ad hoc and sensor networks.

The principle of ALOHA is very simple. As soon as a station generates a frame, it immediately sends it onto the channel. If an acknowledgement is received within a specified time, it indicates that the transmission was successful, otherwise, it retransmits.

The retransmission strategy is that it waits for a random period of time and then retransmits. If there is another collision, it waits for another random period of time until the retransmission is successful.

An obvious disadvantage of this pure ALOHA protocol is that it is very prone to collisions and results in low throughput. We will analyze its throughput and give several improvement forms below.

Throughput analysis of pure ALOHA

Let's discuss how to determine the throughput of an ALOHA system. Figure 8.7 shows the occurrence of collisions in the ALOHA system. In the figure, packet B might collide with packet A from transmission 1 during $t_1 + 2\tau$, where τ is a packet time. Generally, even the collision only occupies a small portion of a packet, the packet might be useless because of the interference and retransmission is required. The collision decreases the transmission efficiency. The larger the load of ALOHA system, the bigger the probability of collisions will be.

Assume that the overall system requirements for the correct arrival rate of packets are λ successful or accepted messages per second. Because of the collisions, some packets will be unsuccessful or rejected. Let's define the total transmission arrival rate is λ_t, which is equal to the reception rate λ plus the rejection ratio, λ_r, i.e.

$$\lambda_t = \lambda + \lambda_r \tag{8.1}$$

Specify the length of each packet as b bits. The average traffic, or throughput, of a channel can be expressed in bits per second

$$\rho' = b\lambda \tag{8.2}$$

The total throughput on a channel can be defined in bits per second

$$G' = b\lambda_t \tag{8.3}$$

The channel capacity (maximum bit rate) is expressed as R bit per second, and the normalized throughput is defined as follows:

$$r = b\lambda / R \tag{8.4}$$

The normalized total load is

$$G = b\lambda_t / R \tag{8.5}$$

The normalized throughput ρ represents the proportion of throughput in the channel capacity ($0 \le \rho \le 1$). The normalized total load G ($0 \le G \le \infty$) represents the ratio of the total load to the channel capacity. Notice that G can be greater than 1.

Define the transfer time for each packet as

$$\tau = b/R \text{ s/packet} \tag{8.6}$$

By substituting eq. (8.6) into eqs. (8.4) and (8.5), we can get:

$$\rho = \lambda\tau \tag{8.7}$$

and

$$G = \lambda_t\tau \tag{8.8}$$

As long as the other users do not transmit packets on the front of it and the back of it in τ time, the user can continuously transmit information. If another user transmits information in front of it in τ time, the tail will conflict with the current transmitted information. If another user transmits information in the back of it in τ time, it will conflict with the tail of the current transmitted information. So each packet needs at least 2τ interval.

The arrival process of irrelevant users in communication system is usually treated as the Poisson process. The probability of K packets arrived in the τ time interval is

$$P(K) = \frac{(\lambda\tau)^K e^{-\lambda\tau}}{K!}, \quad K \ge 0 \tag{8.9}$$

where λ is the average arrival ratio. The transmission of a user in the ALOHA system does not need to consider other users, so this formula is used to calculate the probability when there is exactly $K = 0$ other packet arriving in the 2τ time interval. That is, the probability P_s of correct transmission of information (no collisions) of a user is assumed to be a Poisson process, which is obtained from eq. (8.8) in terms of λ_t and 2τ with $K = 0$:

$$P_s = P(K=0) = \frac{(2\tau\lambda_t)^0 e^{-2\tau\lambda_t}}{0!} = e^{-2\tau\lambda_t} \tag{8.10}$$

In eq. (8.1), the total transmission arrival ratio is defined as λ_t according to the successful part λ and the unsuccessful part λ_r. By definition, the probability of successful packet transmission can be expressed as

$$P_s = \lambda/\lambda_t \tag{8.11}$$

Solving eqs. (8.9) and (8.10) jointly, we can get the following result:

$$\lambda = \lambda_t e^{-2\tau\lambda_t} \tag{8.12}$$

Solving eqs. (8.11), (8.7) and (8.8) jointly, we get:

$$\rho = Ge^{-2G} \tag{8.13}$$

Equation (8.12) describes the relationship between normalized throughput ρ and normalized load G in the channel of ALOHA system. The ρ will increase with the increasing G until a maximum value is reached. Then, the throughput declines because of the increase of collision. The maximum value of ρ is equal to $1/2e = 0.184$, and the value of G is 0.5. We call this access protocol as pure ALOHA. Pure ALOHA channel only utilizes 18% of the resources. The simplicity of control is at the expense of channel capacity.

Slotted-ALOHA (S-ALOHA)

In order to improve throughput of pure ALOHA, slotted ALOHA (S-ALOHA) was proposed. In S-ALOHA, the BS broadcasts a synchronization pulse sequence to all MSs. Similar to pure ALOHA, the length of the package is constant. Packets are required to be transmitted in the slot between synchronization pulses and only at the beginning of the time slot. This simple change halves the collision ratio, because only messages sent simultaneously in the same slot interfere with each other. For S-ALOHA system, the conflict window is reduced from 2τ to τ, resulting in the relationship between normalized throughput and normalized load G as follows:

$$\rho = Ge^{-G} \tag{8.14}$$

where the maximum value of ρ is $1/e = 0.368$, twice that of pure ALOHA.

The retransmit mode of pure ALOHA system needs to be corrected in the S-ALOHA system, so that if the transmitter (MS) receives a NAK, the MS should repeat it after a random integer multiple delay slot.

Reservation ALOHA (R-ALOHA)

Reservation ALOHA (R-ALOHA) scheme is an important improvement of ALOHA protocol. R-ALOHA works in two important modes: non-reservation (static) mode and reservation mode. Suppose that, at the beginning, the system works in the non-reservation (static) mode. The procedure is as follows:

Firstly, a time frame is created and is divided into several small reservation sub-slots. Then, the MS users use these small sub-slots to reserve packet slots. After a user sends the reservation request, it will then listen for confirmation information and slot allocation. Since the control is distributed, all participants are informed of the static format by receiving synchronization pulses from the downlink.

As long as the reservation is made, the system is then switched to the reservation mode. Firstly, the time frame is divided into $M + 1$ time slots. The first M time slots are used for information transmission and the last slot is divided into sub-slots for reservation requests. Since the control is distributed, all users can receive the transmission data from the downlink and they are familiar with the reservation information and time format, so that the confirmation need not contain information other than the first slot location information. The user only sends messages in the slots among the M slots allocated to them. Newly sent reservation requests must be sent to the sub-slots for reservation. When there is no reservation slot, the system returns to non-reservation mode.

8.2.5.2 Carrier-sense multiple access (CSMA)

In the ALOHA access scheme, the transmitter sends the packet as soon as it has a slot or waits for the beginning of a slot, without considering whether the channel is idle or not. CSMA made a modification. A transmitter will firstly *sense* whether the channel is currently occupied by another user (*carrier*) or not. If the channel is occupied, transmission of any user is not allowed. That is why this accessing method is called *CSMA*. Comparing with ALOHA, the channel efficiency of CSMA is much higher since it will not cause collisions if a user is already in transmission.

In CSMA, there are two important parameters: detection delay and propagation delay. *Detection delay* is the time a transmitter (MS) takes to determine whether the channel is currently idle. *Propagation delay* is the time to take for a data packet from the transmitter at MS to the receiver at the BS. Assume the scenario that the first transmitter MS1 finds the channel is idle at time t_1 and then sends off a packet. Suppose that the propagation delay from MS1 to MS2 is τ_p. Another transmitter MS2 detects the channel at $t_1 + \tau_1$, where $\tau_1 < \tau_p$. It means that the packet from MS1 has not arrived at MS2 yet and MS2 also finds the channel is idle and it sends off another packet. Therefore, a collision is caused. This shows the detection delay and propagation delay are important parameters and they should be much smaller than packet duration in order not to cause collisions.

The CSMA algorithm defines the following user actions or responses:

Delay. A user cannot transmit information when the channel is not free.

Transmission. Without delay, the user can transmit until the end of the packet or detect a collision.

Interruption. If a collision is detected, the packet transmission of the user must be interrupted immediately, and a short blocking signal is transmitted by the user to ensure that all the collision users are aware of the collision.

Retransmit. A user must wait for a random delay before attempting to retransmit.

Back off. The delay before the nth retransmit is a random number with uniform distribution between 0 and 2^n–1 ($0 < n \leq 10$).

There are different versions of CSMA implementations. The most popular ones are as follows:

(1) **Nonpersistent CSMA:**

In this version, an MS transmitter who wants to send off a packet senses the channel and operates as follows:

Step 1. If the channel is idle, it sends off a packet.
Step 2. If the channel is busy, the user waits for a random time duration according to a certain delay distribution. At the end of the delay, the user senses the channel again and repeats the above two steps.

(2) **1-Persistent CSMA:**
If a user sends a packet, the protocol is designed not to let the channel to be idle. The user senses the channel and operates as the following process:
Step 1. In case the channel is idle, the user sends off the packet immediately.
Step 2. In case the channel is busy, the user continuously senses the channel until it is idle and then sends off the packet immediately.

Note that in this protocol, collisions occur at multiple users send packets simultaneously.

(3) ***p*-Persistent CSMA:**
In order to decrease the collisions in 1-persistent CSMA and increase the throughput, p-persistent CSMA is proposed to let the start time of a user who sends off a packet to become random. Especially when a user senses the channel is idle, it sends of the packet with a probability of p, and delays the packet time τ with a probability of $1-p$. The selection of p is to decrease the collision probability and keep the idle duration of successive (non-overlapping) very small. The implantation is by dividing the time axis into micro-slots of duration time and sends off a packet at the beginning of a micro-slot. In this protocol, the user who wants to send a packet works as follows:
Step 1. If the channel is idle, the user sends of the packet with a probability of p, and delays the packet time τ with a probability of $1-p$.
Step 2. The user senses the channel at time $t = \tau$; if the channel is still idle, repeat step 1. If collision occurs, retransmission is then arranged according to the preselected transmission delay distribution.
Step 3. The user senses the channel at time $t = \tau$; if the channel is busy, the user will wait until the channel is idle and then repeat steps 1 and 2.

(4) **CSMA with collision detection:**
In this protocol, a node is responsible for observing whether two MS transmitters start to transmit simultaneously. If this event occurs, the transmission is immediately terminated. This method is not common in wireless PR.

(5) **Data-sense multiple access (DSMA):**
This protocol needs a control channel in the downlink to indicate the channel is idle or not, which is implemented by a "busy/available" signal to be transmitted at periodic intervals. If a user who has a packet to transmit finds that the channel is available, it sends off the packet immediately. Note that this protocol is generally implemented in the scenario with a central node (BS), or it would be too costly to design a control channel for each link in a peer-to-peer network.

Like the ALOHA protocol, these CSMA versions can also be implemented in slotted form, which means that each packet must be sent off at the beginning of a slot, thereby decreasing the probability of collisions.

8.3 Cellular system design fundamentals

8.3.1 Channel reuse principle

Channel reuse is fundamental in the cellular system design. The primary motivation is to overcome the major limitations of conventional mobile telephone systems that deploy a high-transmitter power BS in large geographical area to provide services to limited users. When we discuss the channel reuse, we need to also consider the multiple access methods, because the orthogonal multiple access and non-orthogonal multiple access behave differently. In ideal orthogonal multiple access methods, we can ignore intercell interference. The important thing to be considered is the co-channel interference because the same channel will be reused at a certain distance in the cellular system. For CDMA, it is not the case. The spreading codes will be used in every cell. In the downlink, since the codes from all BSs that arrive at a mobile has a timing offset, they are not synchronized, which result in attenuation in autocorrelation at the offset time, and the intercell interference is attenuated accordingly. For wideband CDMA system, a unique non-orthogonal code is assigned to each cell. This unique code is used on the top of the orthogonal codes used in that cell and this reduces the intercell interference by roughly its processing gain. Therefore, we assume the spreading code is used in every cell of a CDMA system, and its reuse distance is one. In the following, we mainly discuss the reuse problem in FDMA and CDMA systems.

8.3.1.1 Reuse distance

Let's discuss the cell and reuse concept. Figure 8.8 shows a geographical area such as a city that is divided into nonoverlapping cells. In this figure, we assume that there are three different channel sets used in this city, c_1, c_2, c_3, and they can be frequency band sets or time slot sets. Each cell in this city is assigned a channel set, and this channel set is reused at certain spatially separated locations. This reuse of channels is called *frequency reuse* or *channel reuse*. The cells that have assigned the same channel set are called *co-channel cells*. A *cluster of cells* can be defined by cells that all use different frequencies. *Cluster size* is then defined as the number of cells in a cluster. In Fig. 8.8, the cluster size $N = 3$. The inverse of cluster size, $1/N$, is defined as the *reuse factor*, since only $1/N$ of the total available channels of the system is assigned to each of the cells within a cluster.

Inside a cluster, there is no co-channel interference. The number of channels as well as the *capacity* is known in a cluster, where the capacity here means the number of users or communication devices that can be supported at the same time. The capacity of the cellular system is actually determined by the cluster size as well as the cell size.

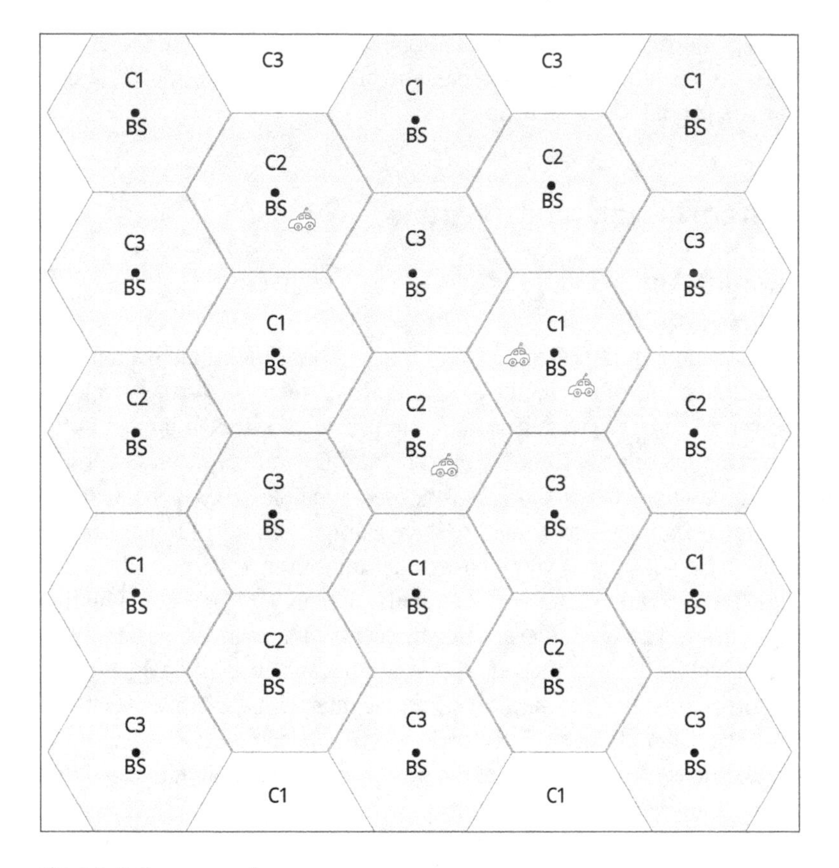

Fig. 8.8: Cell coverage of an area.

A critical question is that how to determine the reused distance? The main consideration is that the performance of the users inside the cell will not be degraded to a non-tolerable level because of the interference from the co-channel cell signals. This is related to the channelization technique, the signal propagation characteristics and the desired acceptable performance for each user. We define the normalized reuse distance between two cells that can use the same channels' reuse distance, $D_{\text{norm}} = D/R$, where R is basically the radius of a cell and might be slightly different because of the shape of cell which will be discussed lately. Apparently, if the cluster size is 1, all the channels will be used in each cell; thus, the capacity gets its maximum value. However, because of the co-channel interference, the signal-to-interference power ratio (SIR) at the boundary of two adjacent cells would be around 0 dB according to the path loss model of the signal from two adjacent BSs. In fact, if the signals from other BSs are considered, the SIR would be less than 0 dB. We know that SIR of 0 dB cannot support reliable communications. In order to guarantee reliable communications, the required SIR for analog FDMA systems (like the first-generation cellular systems' Advanced Mobile Phone Service or Total Access Communication System) is about 18 dB, which corresponds to a cluster size of 21, whereas the required SIR for digital TDMA systems (like second-generation cellular system GSM) is about 10 dB, which corresponds to a cluster size of 7 or less.

8.3.1.2 Determination of the cell shape

How to determine the shape of the cells of a cellular system? According to the path loss model in Chapter 2, the ideal shape of a cell would be circle where the BS is located at the center of it so that the MSs at the boundary of the cell can receive an equal minimum power. Assuming that the background noise and interference is the same, the signal-to-interference plus noise power ratio as well as the system performance at the cell boundary would be equivalent. Of course, because of shadowing, the actual contours of constant received power based on path loss and average random shadowing would be an amoeba-like shape. Figure 8.9 shows the ideal coverage and actual coverage of a cell.

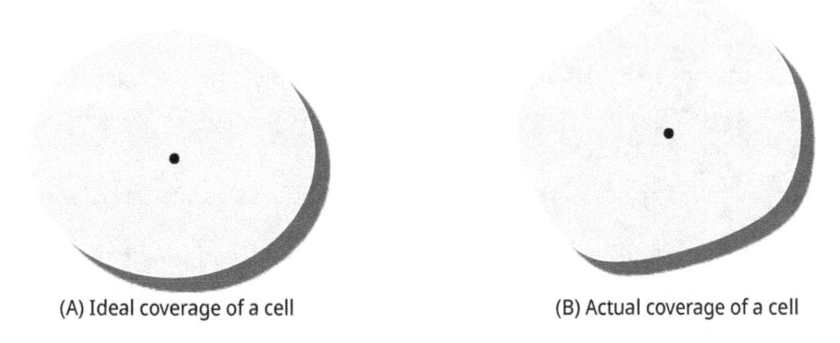

Fig. 8.9: Coverage of a cell.

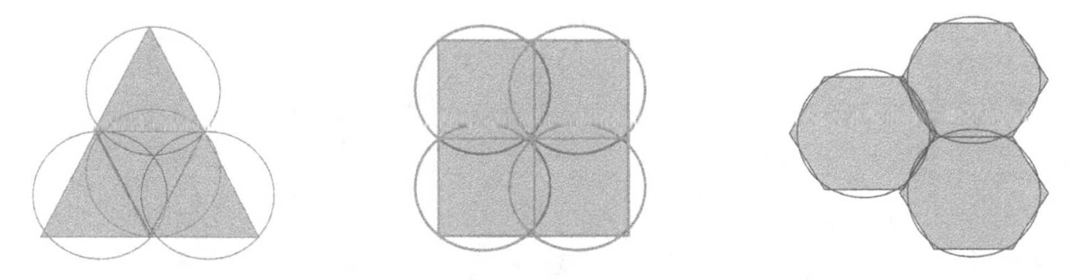

Fig. 8.10: Determination of cell shape.

However, a geographical area cannot be filled by circles without either gaps or overlaps. It can be proved that shapes like triangle, square and hexagon can satisfy the condition of filling a plane area seamlessly and without overlapping, because 360° is an integral multiple of their inner angles. Among the three shapes, hexagon approximates a circle and a large number of hexagons can make up a beehive pattern, and this can be shown in Fig. 8.10. Thus, hexagons are taken as the "basic" cell shape for analysis and theoretical designs. In practice, because of the changing population density and different terrain and huge buildings, practical cell design requires computer simulations or measurements.

8.3.1.3 Channel planning based on hexagonal cells

In the following, we'll discuss the channel planning based on hexagon cells. How to assign channel sets for all the cellular BSs within a system is called *frequency planning* or *channel planning*. As stated above, the capacity of the system is determined by the cluster size and SIR requirements. Figure 8.10 is actually a kind of channel planning, where cells with the same label c_i, $i = 1,2,3$, use the same channel set. Of course, there are only certain cluster sizes, and cell layouts are possible to cover a whole plane and keep the reuse distance constant in a specific cluster size and layout. In order to guarantee that all cells are connected without gaps between adjacent cells, the cluster size N in a hexagon cell system can only be the values that satisfy the following equation:

$$N = i^2 + ij + j^2 \tag{8.15}$$

where i and j are integers greater than or equal to zero. If a channel set is assigned to a given cell, the following procedure is the method to find the nearest co-channel neighbors:

Step 1. Move i cells along any chain of hexagons.

Step 2. Turn 60° counter-clockwise and move j cells.

In the example shown in Fig. 8.10, $N = 3$, $i = 1$ and $j = 1$, probably it is too simple to demonstrate the procedures above. We give two other examples in Fig. 8.11, and the readers may check them by yourself.

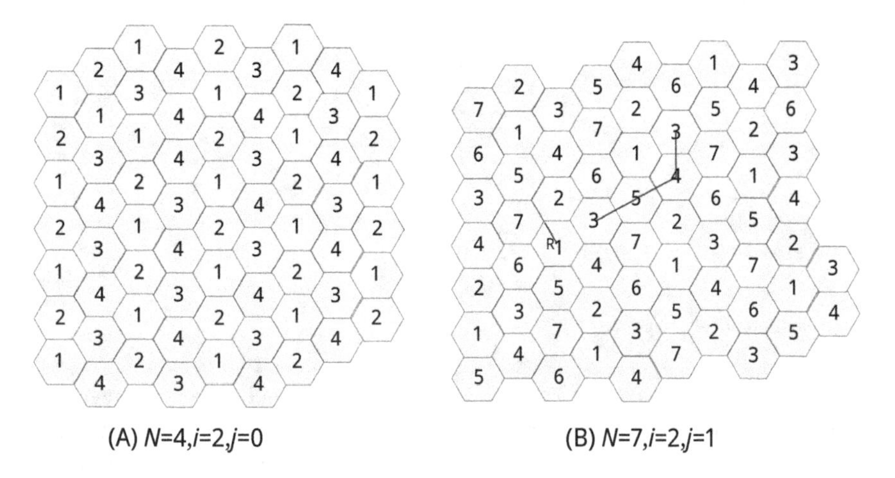

Fig. 8.11: Channel planning examples.

By this procedure, we can also find the reuse distance of a certain channel planning strategy. Let's define the radius of the hexagonal cell (the radius of its outer circle and the line segment from the center to the vertex) is R, the distance of two adjacent cells would be $\sqrt{3}R$, and the distance of two nearest co-channel neighbors by the above two steps can be calculated as

$$D = \sqrt{3}R\sqrt{i^2 + j^2 - 2ij\cos 120^\circ} = \sqrt{3}R\sqrt{i^2 + j^2 + ij} \tag{8.16}$$

Substituting eqs. (8.14) into (8.15), we get the reuse distance of the cellular system based on hexagonal cells:

$$D = \sqrt{3N}R \tag{8.17}$$

where R is the radius of the hexagonal cell and N is the cluster size.

Recall the definition of normalized reuse distance, which is D/R; so we obtain the normalized reuse distance for hexagonal cell clusters:

$$D_{\text{norm}} = \sqrt{3N} \tag{8.18}$$

Table 8.1 lists the typical cluster size and the corresponding normalized reuse distance. Please note that we assume omni-directional antennas are mounted at the BSs.

Tab. 8.1: Typical cluster size and the corresponding normalized reuse distance.

Cluster size	1	3	4	7	9	12	13	16	19	21
Normalized reuse distance	-	3	3.46	4.58	5.2	6	6.24	6.93	7.55	7.94

Example 8.1 Channel planning for a simplified MF-TDMA system.
Suppose in the system each frequency band is 200 kHz wide and can support eight TDMA users.

The required SIR is 10 dB for TDMA. If a fading margin of 10 dB is kept, the mean values of the signal power is required to be at least 20 dB (100) stronger than that of the interference power at the cell boundary. The distance between the desired BS and the endpoint of the radius of the hexagon cell is R, and the distance between the interfering co-channel cell center and the center of the desired cell is (approximately) D-R. Assume that the power falls off with d^{-4}, it is required that

$$\frac{D-R}{R} = (100)^{1/4} = 3.16 \tag{8.19}$$

So that $D/R = 4.16$. By checking Tab. 8.1, the smallest cluster size $N = 7$ can guarantee that $D/R \geq 4.16$. Therefore, the cluster size $N = 7$ can be chosen.

Suppose that an operator is licensed 50 M spectrum, 25 MHz for uplink and 25 MHz for downlink. Each of the uplink and downlink banks can be partitioned into a 200 kHz grid and the outer 100 kHz of each 25 MHz band are not left unused for guard bands. Then there are 124 200 kHz subbands in uplink and downlink. There are about $124/7 \approx 17$ subbands can be used in each cell, and it support $17 \times 8 = 136$ simultaneously users. In practice, because the control channels also take up bandwidth, the actual supported system capacity is smaller than that.

The above discussion shows that the capacity is limited by the cell size. If we shrink the cells or divide one cell into multiple small cells, and keep all aspects of the cellular system scaled so that the SNIR of each user inside a cell remains unchanged, the capacity will be increased. However, propagation properties are related to cell size, and the perfect scale is impossible. Small cell size also increases the probability of handoffs, system cost and management load of the cellular system. Other methods that can increase the capacity can be higher order modulations, better coding methods, MIMO, MUD and so on.

8.3.1.4 Dynamic channel assignment

When the channel planning is finished, each cell can be allocated to a predetermined channel set. If a call within the cell is made, the system will check if there is any unused channel. If there is not any channel available in the cell, the call will be blocked. This fixed channel assignment has drawbacks. There may be such a scenario, and there are still many idle channels in the neighbor cells, but the current cell had to block new calls because there is no channel available. There are some dynamic channel assignment policies to overcome this problem.

One solution is to borrow channels from a neighbor cell if no channel is available. This borrowing must not affect any ongoing calls of the donor cell and this is controlled by the MSC.

In dynamic channel assignment policy, instead of determining the channel set in advance, the channel is allocated each time a call request is made. The channel assignment algorithm is performed in the MSC by considering a series of costs, such as the likelihood of future blocking and the reuse distance, after it receives a request from a BS where the call request is made. The cost of dynamic channel assignment policy is that it has a large storage and computational complexity, because it requires the support of real-time information, such as the channel availability state, the traffic distribution information and the radio signal strength indications of all occupied channels. This increases the storage and computational complexity.

8.3.2 Handoff processing

During a connection period, if an MS moves from one cell to one of its neighbor cells, its serving BS must be changed. This procedure is called *handoff* or *handover*. Handoff should be transparent to the user and be performed as infrequently as possible.

Handover processing is an important task in any cellular network. In many systems, the priority of handover is higher than that of call setup requests when allocating available channel in a cell. The procedure of handoff in different system may be different, but in a similar way as follows:

Step 1. Determine if the handoff condition is satisfied or not.
The system continuously monitors the RSSI of each channel to see if a connection meets the handoff requirements. This measurement and monitoring can be done solely by the BS or measured by the MS and reported to the BS. The system designer must specify the minimum signal level to start the handoff process. The value cannot be too big so that unnecessary handoff does not occur. It also cannot be too small so that the system has enough time to complete the handoff procedure before the connection is lost because of too low signal strength. It should be carefully selected to be a little bit higher than the minimum usable signal for acceptable performance at the BS (for voice communications, the typical value ranges from –90 to –100 dBm). If the measurement is done by the MS, the signal strength from the neighboring BS will also be measured. If the signal received from a neighboring BS begins to surpass the signal received from the current BS at a certain period of time of a certain level, the handoff process should be started. It is also necessary to ensure that the signal drop of the channel of an MS is not due to a temporary signal fading but an actual moving away from its serving BS. This is done by averaging the signal strength of a certain period. The changing speed of the average signal strength should also be monitored so that the MSC makes the decision how fast the handoff must be started.

Step 2. New channel preparation.
A connection is to handoff to a new BS, the first thing is that a new channel must be prepared for the connection at the coming of new serving BS. This may occur in several scenarios. If the new serving BS and the current BS belong to one BS controller (BSC), the handoff is controlled by the controller. If they do not belong to the same BSC, but belong to the same MSC, the handoff is coordinated by the MSC. If they belong to two different MSCs, things would be little bit different. For example, this handoff may change a local call to a roamed call. Anyway, the new controller, BSC, the current MSC or the new MSC that the new serving BS belongs to will inform the new BS to prepare a new channel by the channel assignment techniques we discussed above. If there is no channel available, the handoff would fail.

Step 3. Feed the new channel information back to the MS.
The prepared channel information (carrier frequency, time slot or frequency plus time slot depends on the multiple access strategy) needs feedback to the MS that requires a handoff from the control channel. According to the different cases stated in step 2, this feedback may have different sub-steps. Generally, the BSC informs the new BS transceiver (BST) that prepared the channel. If the handoff occurs in the same BTC, the BTC informs the MS the channel information. If they are not in the same BTC but in the same MSC, this information is first fed back to MSC, then to the original BTC, then to BTS and finally to the MS. If they belong to two MSCs, the information also needs to be transmitted from the new MSC to the original MSC.

Step 4. The MS switches to the new channel and sends handover access bursts. For different systems, the bursts may be different. For example, in GMS, these bursts are shorter than normal ones, because the new BTS does not know the necessary timing advance and has to evaluate it at this time.

Step 5. Through the control channel of the new channel, the MS gets the necessary timing advance and power control information from the new BST.

Step 6. End of the handoff.
After the MS received enough information for communication at the new BS, it tells the MSC that the handoff was successful. By a similar way to step 1, this information will be finally transmitted back the old BS, and the old connection is deleted.

In CDMA-based system, all cells use the same frequency band; thus, an MS can setup connections with two BSs simultaneously. When an MS moves to the boundary of a cell, it may send and receive signals to and from two or more BSs. We know that from Chapter 6, these signals from two or more BSs can be added coherently by the RAKE receiver, and this is a kind of diversity. Suppose an MS moves from cell A to cell B, the signal from cell A is becoming weaker, and that from cell B is becoming stronger. When the signal from cell B is sufficiently strong, the system would make a decision to drop the connection with cell A. This procedure is called *soft handoff*. By contrast, the handoff used in TDMA or FDMA is regarded as *hard handoff*.

8.3.3 A simplified prototype cellular system

In the development of wireless communications for more than 100 years, the concept of cellular is definitely one of the most influential innovations. Wireless communication has become the most dynamic and competitive branch of the communication industry, in which cellular systems make the greatest contribution. Cellular system is by far the most successful and profitable system in wireless communication, and the practitioners of cellular systems have become the most challenging and attractive profession. People are full of enthusiasm and expectation for the new generation of cellular communication system which is full of vigor and renewal. In recent years, the relatively professional high-tech telecommunications have been brought to the public, and the ordinary people have been made to realize the great charm, rich returns and broad application prospects of this new technology. Since the 1980s, the cellular communication system has been developed at the speed of one generation in about 10 years. Up to now, five generations have been developed, and the fifth mobile communication system has begun to enter the commercial stage, while some earlier systems, such as 2G and 3G systems, began to withdraw in some cities.

From 1G to 5G, the system becomes more and more complicated, the bandwidth is wider and wider, the user data rate is higher and higher and the applications are extended from the connection between people to the Internet of things. The services supported are more abundant, and the bearer network has undergone tremendous changes. We intend to take a simple prototype network as an example to describe the cellular network model, leaving aside the specified and complicated network.

8.3.3.1 Architecture and assumptions of the simplified cellular prototype

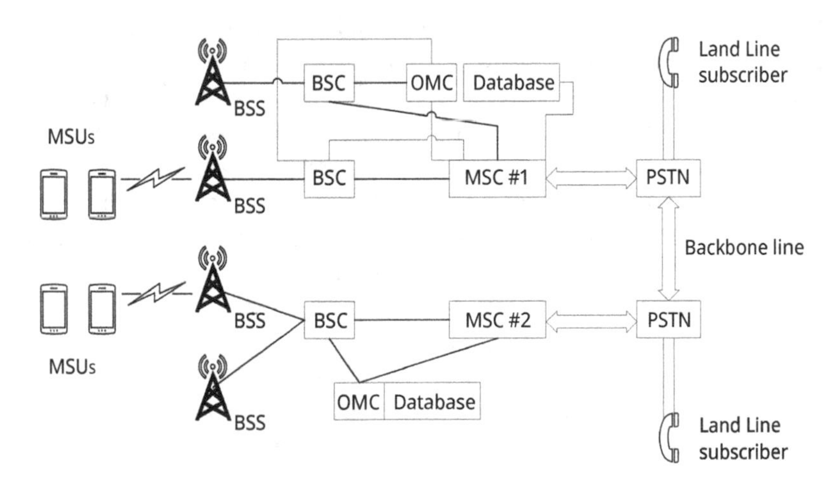

Fig. 8.12: Simplified cellular system architecture.

Figure 8.12 shows such a simplified cellular network. This figure shows two MSCs, which might be corresponding to two cities. For simplification, each MSC only connects one or two BSCs, and each BSC connects to one or two BTSs. We assume that the channel planning and channel assignment has been implemented in some method we have discussed in Section 8.3.1. In the downlink of the BS, a small number of common control channels are used to send BS information. The MSs rapidly scan this control channel to determine the channel with maximum signal strength at any time and tuned to this strongest control channel. The same singling and control data are also broadcasted over all the control channels so that all MSs inside a cell can receive information from the cellular system; thus, all calls from public switched telephone network (PSTN) can be routed to the desired MS.

8.3.3.2 Main parts of the simplified cellular prototype

The prototype consists of mainly four parts: the MS unit (MSU), the BS subsystem (BSS), the MSC and the operations and maintenance center (OMC). In this section, we use the notations consistent with the existing mobile communication systems such as the GSM (Mouly et al. 1992, Molisch 2011, Singal 2011), so that it would be easy for readers to understand the actual systems in their further work and study.

Mobile station unit (MSU)

MSU basically consists of two parts: the cellular phone and the subscriber identifier. These parts may have specific names in different standards. For example, in GSM, the mobile phone is also called the mobile equipment (ME), which includes an antenna, a transceiver and a microprocessor. These components cooperate with each other to accomplish various functions of mobile phones. Each GSM mobile phone sold through regular channels in the world has a unique International Mobile Equipment Identity (IMEI) code. IMEI codes are planned by the GSMA and authorized to be distributed by regional organizations. In China, the Telecommunication Terminal Test Technology Association of the Ministry of Industry and Information Technology is responsible for the authentication of domestic mobile phones. Other distribution agencies include BABT (British approvable Board of Telecommunication) of the United Kingdom and CTIA (Cellular Telephone Industries Association) of the United States. CDMA mobile phone uses Mobile Equipment Identifier code, which is different from the IMEI code. The subscriber identifier is called Subscribe Identify Module (SIM), which is an electronic smart card. SIM consists of CPU, ROM, RAM, EEPROM and I/O circuits. SIM identifies the subscriber and the corresponding information relative to the services. When the SIM is plugged into the ME, the ME can send commands to the SIM card. The SIM card should be executed or rejected according to the standard specifications.

Base station subsystem (BSS)

The BSS consists of two parts: base transceiver stations (BTSs) and the BSCs. The two parts may be co-located or connected by wired or wireless connections. The BTS consists of transceivers and antennas. The antennas are generally mounted on a tall tower. The interface between BSS and the MSs is called *air interface*. The air face is specified by the standard so that mobile phones made by different manufacturers can access the system. The BSC is responsible for the control functions, such as channel assignment, maintenance of link quality, power control, coding and encryption. If a handoff occurs between two BTSs that are controlled by one BSC, the handoff is also coordinated by the BSC, as discussed in Section 8.3.2.

Mobile switching center (MSC)

The MSC hosts the central control functionality of the system. Its functions include the traffic control between different MSCs, the interface with the *Public Service Telephone Network* (PSTN) or other networks, call processing, mobility management, paging and location update. During call processing, it may need to access the database so that certain functions such as call processing and authentication can be performed. The database contains all subscriber information of this MSC, and also contains all the information about mobile subscribers who come from other MSCs and is currently roaming in the network of this MSC with permission. When a subscriber requests a call, it must be verified if the call is allowed or not.

Operations and maintenance center (OMC)

The OMC is responsible for operation organization and maintenance. These functions include billing activities, handling when software and hardware malfunctions occur, software upgrading, identifying and blocking the further functions of MSU that causes interference beyond regulation and collecting data about traffic as well as the quality of the links.

8.3.3.3 Prototype cellular system operations

According to the duplex mode, a call over the cellular systems requires downlink and uplink channels. In a connection between two subscribers, besides the voice or data transmitted which is carried by the traffic channel, there are also many control information in the setup, termination and during the connection. The control information is carried by the control channel. Even in the non-connection period, the MS and BS should periodically maintain information exchange to ensure that the network knows the subscriber's status and the subscriber knows which BS it belongs to. These information exchanges are also through control channels. In this section, we give typical cellular system operations, such as MS registration, establishing a connection and termination of a connection. The handoff is not given here; the reader can refer the general steps in Section 8.3.2 and give specific procedures in different cases.

Mobile station registration

In the last section, we discussed about SIM card. A subscriber is issued the SIM card by a certain Mobile Network Operator, and a valid SIM card corresponding to two identity numbers. The most familiar one to public is the *MS ISDN Number*, which is usually called phone number. The second code is the *International Mobile Subscriber Identity* (IMSI), and it is also a unique identification code to be used to distinguish different users in cellular networks. The MS sends the IMSI to the network in a 64-bit field in certain procedure. In operation, a subscriber also corresponds to other identity numbers that are saved in the SIM card. For example, in order to avoid being identified and tracked by listeners, in most cases, the randomly generated Temporary Mobile Subscriber Identity is used to replace IMSI in the communication between MSs and networks. When a subscriber is roaming, it also has an *MS Roaming Number* (MSRN) which is used for routing of connections. The SIM card also saves other information such as the authentication key, and the access rights which is the permanent security information. It also saves the location area during the operation.

An active subscriber is registered to one MSC of a network. This MSC is the home location for the subscriber so that the Home Location Register (HLR) of the MSC keeps the IMSI and related information of the subscriber. When the subscriber is roaming into the area of other MSC as a visitor and authorized to use the network, its related information is stored into the Visitor Location Register (VLR) of the MSC it roamed into.

All BSs within a network continuously broadcast control information in the downlink control channel. When a cellular phone with a valid SIM card is switched on, it first scans the control channel in specified frequency bands and decides which one is from the nearest Base Transceiver Station (BTS) from the signal strength. It then continuously monitors the signal, and restarts the scanning process to find the strongest signal when the signal drops below a certain threshold. By this way, a mobile subscriber gets registered with a BTS as the current active BTS is connected. Of course, authorization should be performed by the MSC to see if the subscriber has the right to be registered to the MSC. The MSC can track the location of the subscriber by paging it in the control channel. If this MSC is the not the home MSC of the subscriber, the related information is stored in the VLR and fed the location information back to its home MSC so that when a call is made to the subscriber, it can be found through its home MSC.

Establishing a connection

Establishing a connection includes several different scenarios, we list some of them as follows:

A connection from a mobile subscriber to a landline.

A connection from a mobile subscriber to another mobile subscriber within the same cell, in a different cell within the same MSC or in another cell in a different MSC.

Or a connection from a landline phone to a mobile subscriber.

In the following, we pick a connection from a mobile subscriber to another mobile subscriber in another cell but within the same home MSC as an example. Interested readers can think about the differences in other scenarios and give supplements.

A connection setup procedure from a mobile subscriber to mobile subscriber includes the following steps:

Step 1. The mobile subscriber initializes the connection by inputting the phone number of the callee in the keyboard, and then presses the send button. By this way the subscriber requests a dedicated control channel through a random access channel. The BS grants the mobile subscriber a dedicated control channel by a specific grant channel, and then the mobile subscriber sends the callee's phone number (MS ISDN) and its own identification number (IMSI) and related information to the BSC. The BSC sends this information to MSC to check whether the subscriber is allowed to make the request or not. Then the MSC marks this subscriber as busy so that the incoming calls to this subscriber will be blocked.

Step 2. The MSC lets the BSC to allocate a free traffic channel to the subscriber, and this information is forwarded to the BTS and the mobile subscriber.

Step 3. The MSC knows callee's home MSC is itself by checking the phone number. It checks in the HLR for the subscriber's information about the current location area of the subscriber and so on.

Step 4. If the MS is roaming, the HLR knows the MSC it is connected to, and sends a request to the MSC that is currently hosting the subscriber. The hosting MSC sends related information such as the MSRN which can be used for routing of this connection. Since the MSRN contains an identification number of the hosting MSC, the MSC would forward the call to the hosting MSC.

Step 5. The MSC that hosts the callee subscriber sends the paging command to all BTSs to locate the callee subscriber via paging channel. Once the location of the callee subscriber is determined, the BSC sends a paging request to check whether it is available or not.

If the callee subscriber is available, it requests a dedicated control channel, and the BSC grants it a dedicated control channel. By using the dedicated control channel, the MSC lets the BSC to allocate a free traffic channel to the subscriber. So far, the connection is established.

Termination of a connection

Either the caller or the callee subscriber (mobile or landline) in a connection can terminate the call. If a mobile subscriber terminates the connection, it sends this request via a control channel to the BTS, then to MSC, the voice channel is released and this information is forwarded to the MSC hosting another subscriber (for a mobile subscriber) or the PSTN (for landline) to release the related traffic channel. The mobile subscriber returns to MS registration procedure. The landline can also terminate the connection in a similar way. There is also another case that the SNR is too low because of the weakness of the signal or too strong the interference so that the connection had to be terminated and the MSC is informed and the related resource is released. We call this situation as *call drop*.

Problems

8.1 What are the advantages of TDD over FDD, and what is the disadvantage?
8.2 What scenario is FDMA suitable for and why?
8.3 How to overcome the near-far problem of CDMA?
8.4 Suppose the SIR required for an FDMA system is 18 dB, and a 15 dB fading margin is needed to guarantee the performance, what cluster size should be chosen?
8.5 A simplified cellular system is shown in Fig. 8.12; if we number the four cells in figure 1, 2, 3 and 4 from top to the bottom, give out the handoff process for a subscriber to move from cell 2 to cell 3.
8.6 Give the steps of establishing a connection from landline to a cellular phone.

References

Andersen et al. 1995. Andersen JB, Rappaport TS, Yoshida S. Propagation measurements and models for wireless communications channels. IEEE Commun. Mag., 1995, 33(1): 42–49.

Andrews et al. 2001. Andrews MR, Mitra PP, De. Carvalho R. Tripling the capacity of wireless communications using electromagnetic polarization. Nature, 2001, 409: 316–318.

Ayanoglu et al. 1987. Ayanoglu E, Gray RM. The design of joint source and channel trellis waveform coders. IEEE Trans. Inform. Theory, 33(6): 855–865, 1987.

Bello et al. 1962. Bello PA, Nelin BD. The influence of fading spectrum on the bit error probabilities of incoherent and differentially coherent matched filter receivers. IEEE Trans. Commun. Syst., June 1962, 10(2): 160–168.

Berger et al. 1975. Berger T, Davisson LD. Advances in Source Coding. Vienna, Austria, Springer, 1975.

Blaunstein et al. 2007. Blaunstein N, Christodoulou C. Radio Propagation and Adaptive Antennas for Wireless Communication Links: Terrestrial, Atmospheric and Ionospheric. Hoboken, NJ, USA, John Wiley & Sons, Inc, 2007.

Boccuzzi 2008. Boccuzzi Joseph. Signal processing for wireless communications, New York , NY, USA, McGraw-Hill Education, Jan., 2008.

Bottomley et al. 2000. Bottomley GE, Ottosson T, Wang YPE. A generalized RAKE receiver for interference suppression. IEEE J. Sel. Area. Commun., 2000, 18: 1536–1545.

Chuang 1987. Chuang J. The effects of time delay spread on portable radio communications channels with digital modulation. IEEE J. Selected Areas Commun., June 1987, SAC-5(5): 879–889.

Cimini 1985. Cimini LJ. Analysis and simulation of a digital mobile channel using orthogonal frequency division multiplexing. IEEE Trans. Commun., 1985, 33: 665–675.

Cover et al. 2012. Cover TM, Thomas JA. Elements of Information Theory (2nd Edition), Hoboken, NJ, USA, Wiley, 2012.

Cox et al. 1983. Cox DC, Murray RR, Norris AW. Measurements of 800MHz radio transmission into building with metallic walls. Bell Syst. Tech. J., Nov. 1983, 62(9): 2695–2717.

De Toledo et al. 1998. De Toledo AF, Turkmani AMD. Estimating coverage of radio transmission into and within buildings at 900, 1800, and 2300 MHz. IEEE personal communications. 1998, 5(2): 40–47.

Durgin et al. 1998 Durgin G, Rappaport TS, Xu H. Partition-based path loss analysis for in-home and residential areas at 5.85 GHz. IEEE GLOBECOM 1998 (Cat. NO. 98CH36250). Vol. 2: 904–909, Piscataway, NJ, IEEE, 1998.

Foschini 1996. Foschini GJ. Layered space-time architecture for wireless communication in fading environments when using multi-element antennas. Bell Labs Techn. J., 1(2): 41–59, Autumn 1996.

Foschini et al. 1998. Foschini GJ, Gans M. On limits of wireless communications in a fading environment when using multiple antennas, Wireless Pers. Commun., March 1998, 6: 311–355.

Gallagher 1961. Gallagher. Low Density Parity Check Codes, Ph.D. thesis, Massachusetts Institute of Technology (1961).

Gray 1990. Gray RM. Source coding theory, Norwell, MA, USA, Kluwer Academic Publishers, 1990.

Goldsmith et al. 1998. Goldsmith AJ, Effros M. Joint design of fixed-rate source codes and multiresolution channel codes. IEEE Trans. Commun., 46(10): 1301–1312. Oct. 1998.

Goldsmith 2005. Goldsmith AJ. Wireless Communications. New York, NY, USA, Cambridge University Press, 2005.

Goyal et al. 2015. Goyal S, Liu P, Panwar S, Difazio R, Yang R, Bala E. Full duplex cellular systems: Will doubling interference prevent doubling capacity?. IEEE Commun. Mag., May 2015, 53(5): 121–127.

Gurunathan et al. 1992 Gurunathan S, Feher K. Multipath simulation models for mobile radio channels. Vehicular Technology Society 42nd VTS Conference-Frontiers of Technology, Piscataway, NJ, IEEE, 1992: 131–134, May 1992.

https://doi.org/10.1515/9783110751437-009

Hoppe et al. 1999. Hoppe RG, Wölfle, and. Landstorfer FM. Measurement of building penetration loss and propagation models for radio transmission into buildings, *Proc. IEEE Vehicular Technology Conference*, pp. 2298–2302, Piscataway, NJ, IEEE, 1999.

Jakes 1974. Jakes WC. Microwave Mobile Communications. Wiley, New York, 1974. Reprinted by IEEE Press.

Kam 1998. Kam PY. Tight bounds on the bit-error probabilities of 2DPSK and 4DPSK in nonselective Rician fading. IEEE Trans. Commun., 46(7): 860–862, July 1998.

Kohno et al. 1995. Kohno R, Meidan R, Milstein LB. Spread spectrum access methods for wireless communications. IEEE Commun. Mag., 33(1): 58–67, Jan. 1995.

Kraemer et al. 2009. Kraemer R, Marcos DK. Short-Range Wireless Communications: Emerging Technologies and Applications. Chichester, West Sussex, UK, John Wiley & Sons Ltd, 2009.

Lee 1982. Lee WCY. Mobile Communications Engineering. McGraw-Hill, New York, 1982.

LIU et al. 2005. Guangyi LIU, ZHANG J, ZHANG P. Further vision on TD-SCDMA evolution. In Proceeding of the 2005 Asia-Pacific Conference on Communications, Perth, Western Australia, 3–5 Oct., 2005, 143–147.

Liu et al. 2015. Liu G, Richard YF, Ji H, Victor C M L, Li X. In-band full-duplex relaying: A survey, research issues and challenges. IEEE Commun. Surveys Tutorials, 2015, 17(2): 500–524.

Li et al. 2005. Xiaowei L, Fangjiong C, Gang W. A joint detection method for horizontal-BLAST architecture. In 2005 IEEE International Symposium on Microwave, Antenna, Propagation and EMC Technologies for Wireless Communications, 2005, Vol. 2. 987–990.

MacDonald 1979. MacDonald VH. The Cellular Concept. Bell Syst. Tech. J., January, 1979, 58(1): 15–43.

Molisch 2011. Molisch AF. Wireless Communications (2nd edition). Chichester, West Sussex, John Wiley & Sons Ltd, 2011.

Mouly et al. 1992. Mouly M, MB Pautet. The GSM System for Mobile Communications. Telecom Publishing, 1992.

Okumura et al. 1968. Okumura Y, Ohmori E, Kawano T, Fukuda K. Field strength and its variability in VHF and UHF land mobile services. Rev. Elec. Commun. Lab., 1968, 16: 825–873.

Parsons 1992. Parsons D. The Mobile Radio Propagation Channel. Halsted Press (Division of Wiley), New York, 1992.

Paulraj et al. 2003. Paulraj A, Nabar R, Gore DA. Introduction to Space-Time Wireless Communications. Cambridge University Press, Cambridge, England, 2003.

Paulraj et al. 2004. Paulraj AJ, Gore DA, Nabar RU, Bolcskei H. An Overview of MIMO Communications: A key to Gigabit Wireless. Proc IEEE, 2004, 92(2): 198–218.

Peled et al. 1980. Peled R, Ruiz A. Frequency domain data transmission using reduced computational complexity algorithms. In Procedure IEEE International Conference Acoustics, Speech, and Signal Processing, 1980, Denver, CO, 964–967.

Prasad et al. 2002. Prasad N, Varanasi MK. Optimizing the performance of D-BLAST lattice codes for MIMO fading channels. In 2002 IEEE International Conference on Personal Wireless Communications, 2002, 61–65.

Proakis 2000. Proakis JG. Digital Communications (Fifth Edition). McGraw Hill, 2000.

Ramsey 1970. Ramsey JL. Realization of optimum interleavers. IEEE Trans. Inform. Theory, 1970, 16(3): 338–345.

Rappaport 2002. Rappaport TS. Wireless Communications: Principles and Practice (2nd edition), Upper Saddle River, N.J, Prentice Hall PTR, 2002.

Sabharwal et al. 2014. Sabharwal A, Schniter P, Guo D, Bliss DW, Rangarajan S, Wichman R. In-band full-duplex wireless: Challenges and opportunities. 1EEE J. Sel. Areas Commun., Sep.2014, 32(9): 1637–1652.

Sari et al. 1995. Sari H, Karam G, Jeanclaude I. Transmission techniques for digital terrestrial TV broadcasting. IEEE Commun. Mag., Feb. 1995, 33(2): 100–109.

Scholtz 1982. Scholtz RA. The origins of spread spectrum communications. IEEE Trans. Commun., 1982, 30: 822–854.

Schwartz et al. 1966. Schwartz M, Bennett WR, Stein S. Communication Systems and Techniques. McGraw Hill, New York, 1966. Re-issued by IEEE Press, 1997.

Simon et al. 2000. Simon MK, Alouini M-S. Digital Communication over Fading Channels A Unified Approach to Performance Analysis. New York, NY, USA, Wiley, 2000.

Singal 2011. Singal TL. Wireless Communications. New Delhi, India, Tata McGraw Hill, 2011.

Sklar 1997a. Sklar B. Rayleigh fading channels in mobile digital communication systems part I: Characterization. IEEE Commun. Mag., July 1997, 35(7): 90–100.

Sklar 1997b. Sklar B. Rayleigh fading channels in mobile digital communication systems part II: Mitigation. IEEE Commun. Mag., July 1997, 35(7): 102–109.

Stutzman et al. 1997. Stutzman WL, Thiele GA. Antenna Theory and Design (2nd edition). Wiley, New York,1997.

Stüber 2002. Stüber GL. Principles of Mobile Communication, New York, NY, USA, Kluwer Academic Publishers, 2002.

Telatar 1995. Telatar E. Capacity of Multi-antenna Gaussian Channels. AT&T-Bell Labs Internal Memo, 10(6): 585–595, June 1995.

Telatar 1999. Telatar E. Capacity of multi-antenna Gaussian channels. European Trans. Telecomm. ETT, Nov. 1999, 10: 585–596.

Tse et al. 2005. Tse D, Viswanath P. Fundamentals of Wireless Communications. New York, NY, USA, Cambridge University Press, 2005.

Ungerboeck 1982. Ungerboeck, Channel coding with multilevel/phase signals. IEEE Trans. Inform. Theory, 1982, 28, 55–67.

Viterbi 1967. Viterbi AJ. Error bounds for convolutional codes and asymptotically optimum decoding algorithm. IEEE Trans. Inform. Theory, 13(2): 260–269, 1967.

Viterbi 1965. Viterbi AJ. CDMA – Principles of Spread Spectrum Communication, Addison-Wesley Wireless Communications Series,1995.

Vuceti et al. 2000. Vucetic B, Yuan J. Turbo Codes Principles and Applications. New York, NY, USA, Kluwer Academic Publishers, 2000.

Weinstein et al. 1971. Weinstein SB, Ebert PM. Data transmission for frequency-division multiplexing using the discrete fourier transform. IEEE Trans. Commun. Tech., Oct. 1971, 19(5): 628–634. An earlier publication was J. Salz and S. B. Weinstein, Fourier Transform Communication System, Proc. ACM Conf. Comp.and Commun., Pine Mountain, GA, Oct. 1969.

Weinstein 2009. Weinstein SB. The history of orthogonal frequency-division multiplexing. IEEE Commun. Mag., November 2009, 26–35.

Winters 1984. Winters JH. Optimum combining in digital mobile radio with cochannel interference. IEEE J. Sel. Areas Commun., 1984, 2: 528–539.

Winters 1987. Winters J. On the capacity of radio communication systems with diversity in Rayleigh fading environment. IEEE J. Sel. Areas Commun., June 1987, 5: 871–878.

Wolniansky et al. 1998. Wolniansky PW, Foschini GJ, Golden GD, Valenzuela RA. V-BLAST: An architecture for realizing very high data rates over the rich-scattering wireless channel. In 1998 URSI International Symposium on Signals, Systems, and Electronics. Conference Proceedings (Cat. No.98EX167), 295–230.

Yacoub 1993. Yacoub M. Foundations of Mobile Radio Engineering. Boca Raton, Florida, USA, CRC Press, 1993.

ZHANG et al. 2005. Ping ZHANG, Xiaofeng TAO, Jianhua ZHANG, Ying WANG, Lihua LI, Yong WANG. The Visions from FuTURE:Beyond 3G TDD. IEEE Commun. Mag., January, 2005, 43(1): 38–44.

Zheng et al. 2013. Zheng G, Krikidis I, Li J, Petropulu AP, Ottersten B. Improving physical layer secrecy using full-duplex jamming receivers. IEEE Trans. Signal Process, 2013, 61(20): 4962–4974.

Ziemer et al. 1995. Ziemer RE, Peterson RL, Borth DE. Introduction to Spread Spectrum Communications. Upper Saddle River, NJ, USA, Prentice-Hall, 1995.

Index

https://doi.org/10.1515/9783110751437-010

Also of interest

Artificial Intelligence for Signal Processing and Wireless Communication
Abhinav Sharma, Arpit Jain, Ashwini Kumar Arya and Mangey Ram (eds.), 2022
ISBN 978-3-11-073882-7, e-ISBN (PDF) 978-3-11-073465-2,
e-ISBN (EPUB) 978-3-11-073472-0

Communication, Signal Processing & Information Technology
Faouzi Derbel (Ed.), 2020
ISBN 978-3-11-059120-0, e-ISBN (PDF) 978-3-11-059400-3,
e-ISBN (EPUB) 978-3-11-059206-1

Communication and Power Engineering
R. Rajesh and B. Mathivanan (Eds.), 2016
ISBN 978-3-11-046860-1, e-ISBN (PDF) 978-3-11-046960-8
e-ISBN (EPUB) 978-3-11-046868-7

Volume 3 Machine Learning under Resource Constraints – Applications
Katharina Morik, Jörg Rahnenführer and Christian Wietfeld, 2023
ISBN 978-3-11-078597-5, e-ISBN (PDF) 978-3-11-078598-2
e-ISBN (EPUB) 978-3-11-078614-9

Multiple Access Technologies for 5G:
New Approaches and Insight
Jie Zeng, Xin Su, Bin Ren and Lin Liang (Eds.), 2021
ISBN 978-3-11-066636-6, e-ISBN (PDF) 978-3-11-066636-6
e-ISBN (EPUB) 978-3-11-066597-0

Information and Computer Engineering

Volume 1
Analog Electronic Circuit
Beijia Ning, China Science Publishing & Media Ltd.

Volume 2
Random Signal Analysis
Jie Yang, Congfeng Liu, China Science Publishing & Media Ltd.

Volume 3
Signals and Systems
Baolong Guo, Juanjuan Zhu, China Science Publishing & Media Ltd.

Volume 4
Digital Electronic Circuits: Principles and Practices
Shuqin Lou, Chunling Yang, China Science Publishing & Media Ltd.

Volume 5
Programming in C++: Object Oriented Features
Laxmisha Rai, China Science Publishing & Media Ltd.

Volume 6
Data structures based on linear relations
Xingni Zhou, Zhiyuan Ren, Yanzhuo Ma, Kai Fan, Xiang Ji
Part of the multi-volume work Data Structures and Algorithms Analysis

Volume 7
Volume 2 Data structures based on non-linear relations and data processing methods
Xingni Zhou, Zhiyuan Ren, Yanzhuo Ma, Kai Fan, Xiang Ji
Part of the multi-volume work Data Structures and Algorithms Analysis

Volume 8
Communication Electronic Circuits
Zhiqun Cheng, Guohua Liu, China Science Publishing & Media Ltd.

www.degruyter.com

Printed in the USA
CPSIA information can be obtained
at www.ICGtesting.com
JSHW051914250424
61918JS00008B/20

9 783110 751352